4

Student Solutions Manual

for

Devore and Farnum's Applied Statistics for Engineers and Scientists

Second Edition

Nicholas R. Farnum
California State University Fullerton

THOMSON

BROOKS/COLE

Australia • Canada • Mexico • Singapore • Spain • United Kingdom • United States

Printed in Canada
1 2 3 4 5 6 7   08 07 06 05 04

Printer: Webcom

ISBN: 0-534-46720-2

For more information about our products, contact us at:
Thomson Learning Academic Resource Center
1-800-423-0563

For permission to use material from this text or product, submit a request online at
http://www.thomsonrights.com.
Any additional questions about permissions can be submitted by email to **thomsonrights@thomson.com.**

Thomson Brooks/Cole
10 Davis Drive
Belmont, CA 94002-3098
USA

Asia
Thomson Learning
5 Shenton Way #01-01
UIC Building
Singapore 068808

Australia/New Zealand
Thomson Learning
102 Dodds Street
Southbank, Victoria 3006
Australia

Canada
Nelson
1120 Birchmount Road
Toronto, Ontario M1K 5G4
Canada

Europe/Middle East/South Africa
Thomson Learning
High Holborn House
50/51 Bedford Row
London WC1R 4LR
United Kingdom

Latin America
Thomson Learning
Seneca, 53
Colonia Polanco
11560 Mexico D.F.
Mexico

Spain/Portugal
Paraninfo
Calle/Magallanes, 25
28015 Madrid, Spain

TABLE OF CONTENTS

Chapter 1
Data and Distributions

1.   (a) Minitab generates the following stem-and-leaf display of this data:

```
Stem-and-leaf of C1        N  = 27
Leaf Unit = 0.10

    1      5 9
    6      6 33588
  (11)     7 00234677889
   10      8 127
    7      9 077
    4     10 7
    3     11 368
```

The leftmost column in the Minitab printout shows the cumulative numbers of observations From each stem *to the nearest tail* of the data. For example, the 6 in the second row indicates that there are a total of 6 data points contained in stems 6 and 5. Minitab uses parentheses around 11 in row three to indicate that the **median** (described in Chapter 2, Section 2.1) of the data is contained in this stem. A value close to 8 is representative of this data.

What constitutes large or small variation usually depends on the application at hand, but an often-used rule of thumb is: the variation tends to be large whenever the spread of the data (the difference between the largest and smallest observations) is large compared to a representative value. Here, 'large' means that the percentage is closer to 100% than it is to 0%. For this data, the spread is 11 - 5 = 6, which constitutes 6/8 = .75, or, 75%, of the typical data value of 8. Most researchers would call this a large amount of variation.

(b) The data display is not perfectly symmetric around some middle/representative value. There tends to be some positive skewness in this data.

(c) In Chapter 1, outliers are data points that appear to be *very* different from the pack. Looking at the stem-and-leaf display in part (a), there appear to be no outliers in this data. (Chapter 2 gives a more precise definition of what constitutes an outlier).

(d) From the stem-and-leaf display in part (a), there are 3 leafs associated with the stem of 11, which represent the 3 data values that greater than or equal to 11 (and hence, strict' greater than 10). Therefore, the proportion of data values that exceed 10 is 3/27 = or, about 15%.

3.   A Minitab stem-and-leaf display of this data is:

```
              Stem-and-leaf of C1          N    = 36
              Leaf Unit = .01

                  1        3  1
                  6        3  56678
                 18        4  000112222234
                 18        4  5667888
                 11        5  144
                  8        5  58
                  6        6  2
                  5        6  6678
                  1        7
                  1        7  5
```

Another method of denoting the pairs of stems having equal values is to denote the first stem by L, for 'low', and the second stem by H, for 'high'. Using this notation, the stem-and-leaf display would appear as follows:

```
              3L | 1
              3H | 56678
              4L | 000112222234
              4H | 5667888
              5L | 144
              5H | 58
              6L | 2
              6H | 6678
              7L |
              7H | 5
```

The stem-and-leaf display shows that .45 is a good representative value for the data. In addition, the display is not symmetric and appears to be positively skewed. The spread of the data is .75 - .31 = .44, which is.44/.45 = .978, or about 98% of the typical value of .45. Using the same rule of thumb as in Exercise 1, this constitutes a reasonably large amount of variation in the data. The data value .75 is a possible outlier (the definition of 'outlier' in Section 2.3, shows that .75 could be considered to be a 'mild' outlier).

5.   (a) Two-digit stems would be best. One-digit stems would create a display with only 2 stems, 6 and 7, which would give a display without much detail. Three-digit stems would cause the display to be much too wide with many gaps (stems with no leafs).

2

(b) The stem-and-leaf display below does not give up (truncate) the rightmost digit in the data:

```
64 | 33 35 64 70
65 | 06 26 27 83
66 | 05 14 94
67 | 00 13 45 70 70 90 98
68 | 50 70 73 90
69 | 00 04 27 36
70 | 05 11 22 40 50 51
71 | 05 13 31 65 68 69
72 | 09 80
```

(c) A Minitab stem-and-leaf display of this data appears below. Note that Minitab <u>does</u> truncate the rightmost digit in the data values.

```
 4      64 3367
 8      65 0228
11      66 019
18      67 0147799
(4)     68 5779
18      69 0023
14      70 012455
 8      71 013666
 2      72 08
```

This display tends to be about as informative as the one in part (b). With larger sample sizes, the work involved in creating the display in part (c) would be much less than that required in part (b). In addition, for a larger sample size, the 'full' display in (b) would require a lot of room horizontally on the page to accommodate all the 2-digit leaves.

7. (a)

Number Nonconforming	Frequency	Relative Frequency (Freq/60)
0	7	0.117
1	12	0.200
2	13	0.217
3	14	0.233
4	6	0.100
5	3	0.050
6	3	0.050
7	1	0.017
8	1	0.017

doesn't add exactly to 1 because relative frequencies have been rounded → 1.001

(b) The number of batches with at most 5 nonconforming items is 7+12+13+14+6+3 =55, which is a proportion of 55/60 = .917. The proportion of batches with (strictly) fewer than 5 nonconforming items is 52/60 = .867. Notice that these proportions could

3

been computed by using the relative frequencies: e.g., proportion of batches with 5 or fewer nonconforming items = 1- (.05+.017+.017) = .916; proportion of batches with fewer than 5 nonconforming items = 1 - (.05+.05+.017+.017) = .866.

(c) The following is a Minitab histogram of this data. The center of the histogram is somewhere around 2 or 3 and it shows that there is some positive skewness in the data. Using the rule of thumb in Exercise 1, the histogram also shows that there is a lot of spread/variation in this data.

Relative Frequency

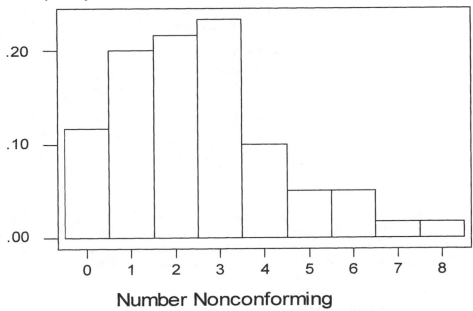

Number Nonconforming

9. (a) From this frequency distribution, the proportion of wafers that contained at least one particle is (100-1)/100 = .99, or 99%. Note that it is much easier to subtract 1 (which is the number of wafers that contain 0 particles) from 100 than it would be to add all the frequencies for 1, 2, 3,… particles. In a similar fashion, the proportion containing at least 5 particles is (100 - 1-2-3-12-11)/100 = 71/100 = .71, or, 71%.

(b) The proportion containing between 5 and 10 particles is (15+18+10+12+4+5)/100 = 64/100 = .64, or 64%. The proportion that contain strictly between 5 and 10 (meaning strictly *more* than 5 and strictly *less* than 10) is (18+10+12+4)/100 = 44/100 = .44, or 44%.

4

(c) The following histogram was constructed using Minitab. The data was entered using the same technique mentioned in the answer to exercise 8(a). The histogram is *almost* symmetric and unimodal; however, it has a few relative maxima (i.e., modes) and has a very slight positive skew.

Relative frequency

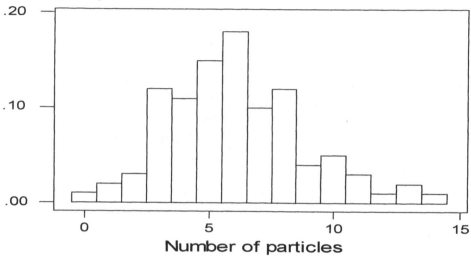

11.   (a) The following stem-and-leaf display was constructed using Minitab:

```
Stem-and-leaf of C1       N  = 47
Leaf Unit = 100

   12      0 123334555599
   23      1 00122234688
  (10)     2 1112344477
   14      3 0113338
    7      4 37
    5      5 23778
```

A typical data value is somewhere in the low 2000's. The display almost unimodal (the stem at 5 would be considered a mode, the stem at 0 another) and has a positive skew.

(b) A histogram of this data, using classes of width 1000 centered at 0, 1000, 2000, ..., 6000 is shown below. The proportion of subdivisions with total length less than 2000 is (12+11)/47 = .489, or 48.9%. Between 200 and 4000, the proportion is (7 + 2)/47 = .191, or 19.1%. The histogram shows the same general shape as depicted by the stem-and-leaf in part (a).

5

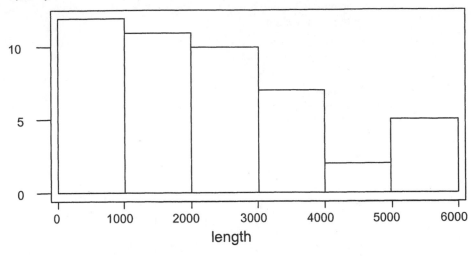

Frequency

13. (a) The frequency distribution is:

Class	Relative Frequency		Class	Relative Frequency
0-< 150	.193		900-<1050	.019
150-< 300	.183		1050-<1200	.029
300-< 450	.251		1200-<1350	.005
450-< 600	.148		1350-<1500	.004
600-< 750	.097		1500-<1650	.001
750-< 900	.066		1650-<1800	.002
			1800-<1950	.002

The relative frequency distribution is almost unimodal and exhibits a large positive skew. The typical middle value is somewhere between 400 and 450, although the skewness makes it difficult to pinpoint more exactly than this.

(b) The proportion of the fire loads less than 600 is .193+.183+.251+.148 = .775. The proportion of loads that are at least 1200 is .005+.004+.001+.002+.002 = .014.

(c) The proportion of loads between 600 and 1200 is 1 - .775 - .014 = .211.

(d) Since 500 and 1000 are not class boundaries, we have to approximate the proportion of loads between these two numbers. For the classes 600-<750 and 750-<900, no approximation is necessary; their accumulated proportions are simply .097 + .066. For the interval 500-<600, which is contained in the class 450-<600, we simply calculate what fraction the distance 600-500= 100 is compared to the entire class width of 600-450 = 150 and take that fraction of the corresponding class proportion.148. In this case, 100/150 = 2/3, so we estimate that about 2/3 of .148, or .0987, is the relative frequency to attribute the the interval 500-<600. Similarly, the interval 900-<1000 should account for about (1000-900)/(1050-900) = 100/150 = 2/3 of the class

6

frequency .019. Therefore, we estimate that the proportion of loads between 500 and 1000 is approximately .0987 + .097 + .066 + .0127 = .274.

15.   (a) A histogram with classes of width 100 appears below. The histogram is positively skewed and a representative data value is around 200.

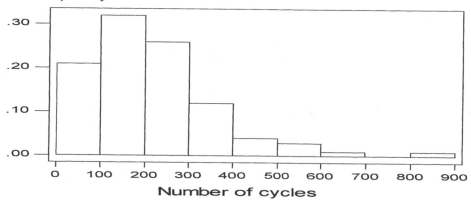

(b) Using class widths of different sizes, as specified in the exercise, the relative frequency histogram would appear as follows:

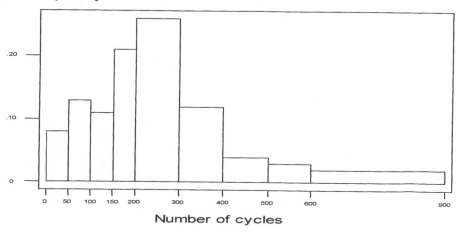

(c) The proportion of data values that equal or exceed 100 cycles is 1-.21 = .79, or 79%.

17.   The histogram of this data appears below. A typical value of the shear strength is around
         5000 lb. The histogram is almost symmetric and approximately bell-shaped.

Relative frequency

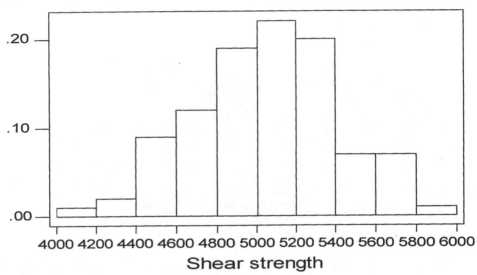

Shear strength

Note → *The following notation will be used to simplify writing out the answers in the*
remainder of this chapter: for example, we will write **Proportion (x > 7)** to mean **"the**
proportion of the x values that exceed 7"; Proportion (3 < x < 7) stands for **"the**
proportion of the x values that lie between 3 and 7", etc.

19.   (a) The density curve forms a rectangle over the interval [4, 6]. For this reason, uniform
         densities are also called **rectangular densities** by some authors. Areas under uniform
         densities are easy to find (i.e., no calculus is needed) since they are just areas of rectangles.
         For example, the total area under this density curve is $\frac{1}{2}(6-4) = 1$.

height = 1/(6-4) = 1/2

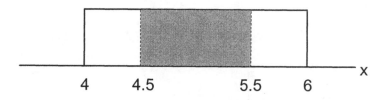

(b) The proportion of values between 4.5 and 5.5 is depicted (shaded) in the diagram below. The area of this rectangle is $\frac{1}{2}(5.5 - 4.5) = .5$. Similarly, the proportion of x values that exceed 4.5 is $\frac{1}{2}(6 - 4.5) = .75$.

(c) The median of this distribution is 5 because exactly half the area under this density sits over the interval [4,5].

(d) Since 'good' processing times are short ones, we need to find the particular value x_0 for which the proportion of the data less than x_0 equals .10. That is, the area under the density to the left of x_0 must equal .10. Therefore, area $= .10 = \frac{1}{2}(x_0 - 4)$, so $x_0 - 4 = .20$ and $x_0 = 4.20$.

21.   (a) The density function is $f(x) = 1/(20\text{-}7.5) = 1/12.5$ over the interval [7.5, 20] and $f(x) = 0$ elsewhere. The proportion of depths less than $k = \frac{1}{12.5}(k - 7.5)$. For $k = 10$, this proportion is $\frac{1}{12.5}(10 - 7.5) = .20$. For $k = 15$ it is $\frac{1}{12.5}(15 - 7.5) = 60$.

(b) The proportion of x values that are at least k is $\frac{1}{12.5}(20 - k)$. The proportion of x values that *strictly* exceed k is also $\frac{1}{12.5}(20 - k)$ because f(x) is a continuous density. For k = 10, this proportion is $\frac{1}{12.5}(20 - 10) = .80$; for k = 15 it is $\frac{1}{12.5}(20 - 15) = .40$.

9

(c) It helps to draw the picture of the density:

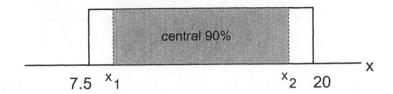

So, the area to the left of x_1 should be .05, i.e., $\frac{1}{12.5}(20-10) = .05$. Similarly, the area to the right of x_2 is also .05, $\frac{1}{12.5}(20-x_2) = .05$. Solving these equations gives $x_1 = 8.125$ and $x_2 = 19.375$.

23.   (a)

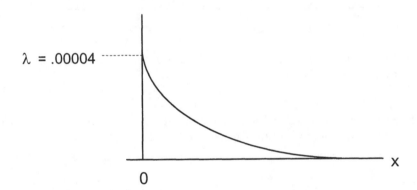

(b) $\int_{20,000}^{\infty} .00004e^{-.00004x}\, dx = \left. \frac{-(.00004)e^{-.00004x}}{(.00004)} \right]_{20,000}^{\infty} = \left. -e^{-.00004x} \right]_{20,000}^{\infty}$

   $= 0 - (-e^{-.00004(20,000)}) = e^{-.8} = .449.$

Note: for any exponential density curve, the area to the right of some fixed constant c always equals $e^{-\lambda c}$, as our integration above shows. That is,

$$\text{Proportion } (x > c) \;=\; \int_{c}^{\infty} \lambda e^{-\lambda x}\, dx \;=\; e^{-\lambda c}.$$

We will use this fact in the remainder of the chapter instead of repeating the same type of integration as in part (a)

Proportion $(x \le 30,000) = 1 - $ Proportion $(x > 30,000) = 1 - e^{-\lambda c} = 1 - e^{-.00004(30,000)} = 1 - e^{-1.2} = .699.$

10

Proportion $(20,000 \leq x \leq 30,000)$ = Proportion $(x > 30,000)$ - Proportion$(x \leq 20,000)$
= $.699 - (1-.449) = .148$.

(c) For the best 1%, the lifetimes must be at least x_0, where Proportion$(x \geq x_0) = .01$, which becomes $e^{-\lambda x_0} = .01$. Taking natural logarithms of both sides, $-\lambda x_0 = \ln(.01)$, so $x_0 = -\ln(.01)/\lambda = 4.60517/.00004 = 115,129.25$. For the worst 1%, we have Proportion$(x \leq x_0)$ $= .01$, which is equivalent to saying that Proportion$(x \geq x_0) = .99$, so $e^{-\lambda x_0} = .99$. Taking logarithms, $-\lambda x_0 = \ln(.99)$, so $x_0 = -\ln(.99)/\lambda = 251.26$.

25. (a) The total area under the density curve must equal 1, so:

$$1 = \int_2^4 c[1-(x-3)^2]dx = c\int_2^4 [1-(x-3)^2]dx = c\left(x-\tfrac{1}{3}(x-3)^2\right)\Big]_2^4 = c[(4-\tfrac{1}{3})-(2+\tfrac{1}{3})] =$$

$\tfrac{4}{3}$ c, so, c = $\tfrac{3}{4}$.

(b) Proportion$(x > 3)$ equals .50. No calculation is needed if you draw a diagram of the density curve and notice that it is symmetric around $x = 3$, which means that exactly half the area under the curve lies on either side of $x = 3$.

(c) Proportion$(3-.25 \leq x \leq 3+.25)$ = Proportion$(2.75 \leq x \leq 3.25)$ =

$$\int_{2.75}^{3.25} f(x)dx = \tfrac{3}{4}\left(x - \frac{(x-3)^2}{3}\right)\Bigg]_{2.75}^{3.25} = .367.$$

27. (a) Proportion$(x \leq 3) = .10 + .15 + .20 + .25 = .70$.
      Proportion$(x < 3)$ = Proportion$(x \leq 2) = .10 + .15 + .20 = .45$.

(b) Proportion$(x \geq 5) = 1 - $ Proportion$(x < 5) = 1 - (.10+.15+.20+.25+.20) = .10$.

(c) Proportion$(2 \leq x \leq 4) = .20 + .25 + .20 = .65$

(d) At least 4 lines will *not* be in use whenever 2 or fewer lines *are* in use. At most 3 lines are in use .45, or 45%, of the time from part (a) of this exercise.

29.	The 10 possible samples of size two are shown below along with the corresponding x values:

sample → {1,2} {1,3} {1,4} {1,5} {2,3} {2,4} {2,5} {3,4} {3,5} {4,5}
   x   →   2   1   1   1   1   1   1   0   0   0

The mass function for x is then:	x → 0 1 2
                                  	$p(x)$ → 3/10 6/10 1/10

31.	(a) Proportion$(z \le 1.78) = .9625$ (Table I, p. 561)

(b) Proportion$(z > .55) = 1 -$ Proportion$(z \le .55) = 1 - .7088 = .2912$.

(b) Proportion$(z > -.80) = 1 -$ Proportion$(z \le -.80) = 1 - .2119 = .7881$.

(c) Proportion$(.21 \le z \le 1.21) =$ Proportion$(z \le 1.21) -$ Proportion$(z \le .21)$
	$= .8869 - .5832 = .3037$.

(d) Proportion$(z \le -2.00$ or $z \ge 2.00) =$ Proportion $(z \le -2.00) + [1-$ Proportion$(z < 2.00)]$
	$= .0228 + [1 - .9772] = .0456$. Alternatively, using the fact that the z density is symmetric around $z = 0$, Proportion$(z \le -2.00) =$ Proportion$(z \ge 2.00)$, so the answer is simply $2 \cdot$Proportion$(z \le -2.00) = 2(.0228) = .0456$.

(e) Proportion$(z \le -4.2)$ $= .0000$

(f) Proportion$(z > 4.33) = .0000$

33.	(a) Let z* denote the value of z that is exceeded by the largest 15% of all z values. Then, Proportion$(z > z^*) = .15$. In terms of left-tail areas, Proportion$(z \le z^*) = .85$. From Table I, Proportion$(z \le 1.04) = .8508$ and Proportion$(z \le 1.03) = .8485$, so, using linear interpolation, $z^* = 1.03 + [(.8500-.8485)/(.8508-.8485)](1.04-1.03) = 1.036$. That is, any z value greater than 1.036 (or approximately 1.04) will be in the top 15%.

(b) Proportion$(z \le z^*) = .25$. From Table I, Proportion$(z \le -.68) = .2483$ and Proportion$(z \le -.67) = .2514$, so interpolation gives
	$z^* = -.67 + [(.2500-.2483)/(.2514-.2483)](-.68-(-.67)) = -.6755$ or, approximately, $z^* = -.675$.

(c) The 4% farthest from 0 consists of the 2% in the upper tail and the 2% in the lower tail. In the upper tail, Proportion$(z > z^*) = .02$, which can be re-expressed as a left-tail area Proportion$(z \le z^*) = .98$. From Table I, Proportion$(z \le 2.05) = .9798$ and Proportion$(z \le 2.06) = .9803$, so $z^* \approx 2.055$. Similarly, a z* value of -2.055 will capture the lower 2% of the z values.

35. (a) Let x = bolt thickness. Then, the statement x ≤ 11 describes bolts that are less than one σ (since σ = 1) to the right of μ = 10. The equivalent area under the z density is z ≤ 1 (since μ = 0 and σ= 1). From Table I, this area/proportion is .8413.

   (b) The region 7.5 ≤ x ≤ 12.5 describes bolts whose lengths are within 2.5 σ's from the mean of 10. The equivalent z curve area is described by -2.5 ≤ z ≤ 2.5. From Table I, this area equals Proportion(z ≤ 2.5) - Proportion(z ≤ -2.5) = .9938 - .0062 = .9876.

   (c) The region x > 11.5 describes bolts that are more than 1.5 σ's above the mean. The corresponding z curve region is z > 1.5. From Table I, Proportion(z > 1.5) = 1 - Proportion(z ≤ 1.5) = 1 - .9332 = .0668.

37. (a) Let x = yield strength. Then, x ≤ 40, when standardized, becomes z ≤ (40-43)/4.5 = - .67. From Table I, Proportion(z ≤ -.67) = .2514. Similarly, standardizing x > 60 yields z > (60-43)/4.5 = 3.78. From Table I, Proportion(z > 3.78) = 1 - Proportion(z ≤ 3.78) = 1- .9999 = .0001.

   (b) Standardizing 40 ≤ x ≤ 50 gives (40-43)/4.5 ≤ z ≤ (50-43)/4.5 or, -.67 ≤ z ≤ 1.56. From Table I, this proportion equals Proportion(z ≤ 1.56) - Proportion(z ≤ -.67) = .9406 - .2514 = .6892.

   (c) We need to find the value of x* for which Proportion(x > x*) = .75, or equivalently, Proportion(x ≤ x*) = .25. The corresponding statement about a z curve is Proportion(z ≤ z*) = .25. From Table I, z* ≈ -.675; i.e., z^* is .675 σ's below the mean. So, x* must be .675 σ's below the mean of the x data: x* = 43 - .675(4.5) = 39.963.

39. (a) Let x = substrate concentration. Standardizing x > .25 yields z > (.25-.30)/.06, or z > -.83. From Table I, Proportion(z > -.83) = 1 - Proportion(z ≤ -.83) = 1 - .2033 = .7967.

   (b) x ≤ .10 standardizes to z ≤ (.10-.30)/.06 = -3.33, so Proportion(z ≤ -3.33) = .0004.

   (c) Let x* be the value exceeded by the largest 5% of x values. Then Proportion(x > x*) = .05. The equivalent statement about the z density is Proportion(z > z^*) = .05, or, in terms of left-tail areas, Proportion(z ≤ z*) = .95. From Table I, z^* = 1.645, which is 1.645 σ's above the mean of the z data. Therefore, x* must be 1.645 σ's above the mean of the x data: x* = .30+ 1.645(.06) = .399.

41. (a) Because x is a discrete variable, the best approximation to the Proportion(20 ≤ x ≤ 40) is to find the area under the x density curve between the points 19.5 and 40.5. That is, Proportion((19.5-25)/5 ≤ z ≤ (40.5-25)/5) = Proportion(-1.1 < z < 3.1) = Proportion(z < 3.1) - Proportion(z < -1.1) = .9990 - .1357 = .8633.

(b) Using the same sort of approximation as in part (a), $\text{Proportion}(x \le 30.5) = \text{Proportion}(z \le (30.5\text{-}25)/5) = \text{Proportion}(z \le 1.1) = .8643$. Similarly, the proportion of reels having *fewer* than 30 flaws is approximated by $\text{Proportion}(x < 29.5) = \text{Proportion}(z \le (29.5\text{-}25)/5) = \text{Proportion}(z \le .9) = .8159$.

43.　(a) Let $x = SO_2$ concentration. Then $\text{Proportion}(x \le 10) = \text{Proportion}(\ln(x) \le \ln(10)) = \text{Proportion}(\ln(x) \le 2.302585)$. Standardizing, this proportion equals $\text{Proportion}(z \le (2.302585\text{-}1.9)/.9) = \text{Proportion}(z \le .45) = .6736$ (from Table I). Similarly, $\text{Proportion}(5 \le x \le 10) = \text{Proportion}(1.609438 \le \ln(x) \le 2.303585) = \text{Proportion}((1.609438\text{-}1.9)/.9 \le z \le (2.302585\text{-}1.9)/.9) = \text{Proportion}(\text{-}.32 \le z \le .45) = \text{Proportion}(z \le .45) - \text{Proportion}(z \le \text{-}.32) = .6736 - .3745 = .2991$.

(b) The constants a and b satisfy the equations $\text{Proportion}(x \le a) = .025$ and $\text{Proportion}(x \ge b) = .025$. Therefore, $.025 = \text{Proportion}(x \le a) = \text{Proportion}(\ln(x) \le \ln(a)) = \text{Proportion}(z \le (\ln(a)\text{-}1.9)/.9)$. From Table I, $\text{Proportion}(z \le \text{-}1.96) = .025$, so $(\ln(a)\text{-}1.9)/.9 = \text{-}1.96$, or, $\ln(a) = 1.9 - 1.96(.9) = .136$. Finally, $a = \exp(.136) = 1.146$. For the other equation, $.025 = \text{Proportion}(x \ge b) = \text{Proportion}(\ln(x) \ge \ln(b)) = \text{Proportion}(z \ge (\ln(b)\text{-}1.9)/.9)$. Using the symmetry of the z density, the right tail area is captured by $z > 1.96$, so $(\ln(b)\text{-}1.9)/.9 = 1.96$, or $\ln(b) = 1.9+1.96(.9) = 3.664$. Therefore, $b = \exp(3.664) = 39.017$.

45. (a) Let x = fracture strength. Then, $\text{Proportion}(x \le 100) = 1\text{-} e^{-(100/125)^5} = .2794$. Similarly, $\text{Proportion}(100 < x < 150) = \text{Proportion}(x > 100) - \text{Proportion}(x > 150) = e^{-(100/125)^5} - e^{-(150/125)^5} = .720594 - .083049 = .6375$.

(b) Let x^* denote the strength that exceeds the weakest 50% of the x values. Then, $.50 = \text{Proportion}(x \le x^*) = 1 - e^{-(x^*/125)^5}$, so $e^{-(x^*/125)^5} = 1\text{-}.50 = .50$. Taking logarithms of both sides, $-(x^*/125)^5 = \ln(.50) = \text{-}.693147$, so $x^* = 125(.693147^{1/5}) = 116.16$.

(c) Let x^* denote the strength that exceeds the weakest 5% of the x values. Then, $.05 = \text{Proportion}(x \le x^*) = 1 - e^{-(x^*/125)^5}$, so $e^{-(x^*/125)^5} = 1\text{-}.05 = .95$. Taking logarithms of both sides, $-(x^*/125)^5 = \ln(.95) = \text{-}.0512933$, so $x^* = 125(.0512933)^{1/5} = 69.01$.

47.   (a) The histogram (below) shows that the data is slightly skewed to the right.

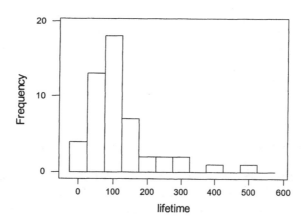

(b) Using a class width of .2, the histogram of the natural logarithms of the data appears below. This histogram appears to be more symmetric and bell-shaped (i.e., normal) than does the histogram of the raw data. This supports the belief that a lognormal distribution provides a reasonable model for the original/raw data.

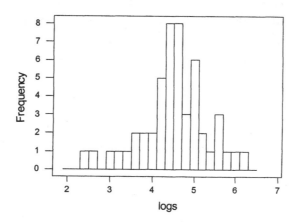

49.   (a) Since X = "hydrocarbon emissions" is lognormal, then Y = ln(X) is normal with mean μ = 4.5 and standard deviation σ = .625. So, Proportion(50 < X < 150) = Proportion(ln(50) < Y < ln(150)) = Proportion(3.91202 < Y < 5.01064). Converting (standardizing) the last expression to the z scale gives Proportion(-.94 < z < .82), which equals (using Table I) .7939 - .1736 = .6203.

    (b) The "best" engines are the ones with the smallest amounts of emissions, so we want to find the value of c for which .01 = Proportion(X < c) = Proportion(Y < ln(c)). From Table I, Proportion(z < -2.33) $\approx$ .01, so ln(c) $\approx$ 4.5 – 2.33(.625) = 3.04375 and therefore c $\approx$ exp(3.04375) = 20.98 g/gal.

51.   (a) Proportion(x $\leq$ 2) = .277 + .365 + .231 = .873 (from Table II, p. 564).

    (b) Proportion(x $\geq$ 5) = .006 + .001 + .000 + ...+ .000 = .007.

    (b) Proportion(x = 0) = .277.

53.   (a) Let x = number of red lights encountered. Then x has a binomial distribution with n = 10, π = .40. Proportion(x $\leq$ 2) = .006 + .040 + .121 = .167 (using Table II, p. 562). Similarly, Proportion(x $\geq$ 5) = .201 + .111 + .042 + .011 + .002 + .000 = .367.

    (b) Proportion(2 $\leq$ x $\leq$ 5) = .121 + .215 + .251 + .201 = .788.

55.   (a) Let x denote the number of defectives in a sample of size 10. Then, x has a binomial distribution with n = 10 and π = proportion of defectives in the batch sampled. A batch is accepted whenever x $\leq$ 2. So, for π = .01, Proportion(x $\leq$ 2) = $\frac{10!}{0!10!}(.01)^0(1-.01)^{10}$ + $\frac{10!}{1!9!}(.01)^1(1-.01)^9 + \frac{10!}{2!8!}(.01)^2(1-.01)^8$ = .90438 + .09135 + .00411 = .9998.

The remaining proportions can be found using Table II, page 562:

For π = .05, Proportion(x $\leq$ 2) = .599 + .315 + .075 = .989
For π = .10, Proportion(x $\leq$ 2) = .349 + .387 + .194 = .930
For π = .20, Proportion(x $\leq$ 2) = .107 + .268 + .302 = .677
For π = .25, Proportion(x $\leq$ 2) = .056 + .188 + .282 = .526

    (b) A graph of the proportion of batches accepted versus π is shown below. The appearance of the graph would be smoother if more points were plotted (i.e., if additional values if π were used).

Proportion of batches accepted

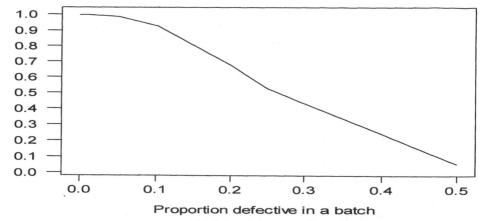

Proportion defective in a batch

Note: For any sampling plan, when $\pi = 0$ the proportion of lots accepted will be exactly 1 because there will be no defectives in the lot, so $x \leq c$ will be true for any nonnegative integer c. So, we can always include the point $(0, 1)$ on an operating characteristic curve.

(c) The acceptance criterion is now $x \leq 1$, so, using results from part (a):

For $\pi = .01$,   Proportion$(x \leq 1) = .90438 + .09135 = .9957$
For $\pi = .05$,   Proportion$(x \leq 1) = .599 + .315 = .914$
For $\pi = .10$,   Proportion$(x \leq 1) = .349 + .387 = .736$
For $\pi = .20$,   Proportion$(x \leq 1) = .107 + .268 = .375$
For $\pi = .25$,   Proportion$(x \leq 1) = .056 + .188 = .244$

The new operating characteristic curve appears below. Because Proportion$(x \leq 1)$ is always less than Proportion$(x \leq 2)$, this plan accepts fewer batches (at any value of π) than the previous plan. In this sense, the plan that accepts batches with $x \leq 1$ is said to be *tighter* (i.e., more discriminating) than the plan that accepts batches with $x \leq 2$. Finding the right plan is a compromise between a plan that is too tight (rejecting too many batches, including ones with low proportions of defectives) and plans that are too loose (which accept too many batches with large proportions of defectives).

Proportion of batches accepted

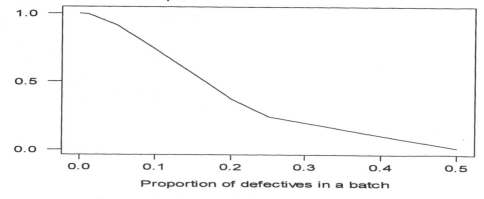

Proportion of defectives in a batch

57. (a) Proportion($x = 1$) = .164 (Table III, p. 565, $\lambda = .2$)

(b) Proportion($x \geq 2$) = 1 - Proportion($x \leq 1$) = 1 - (.819 + .164) = 1 - .983 = .017.

(c) Let y denote the number of missing pulses on 10 disks. Then, Proportion($y \geq 2$) =

1 - Proportion($y \leq 1$) = 1 - (.135 + .271) = .594 (Table III, with $\lambda = 2$).

59. Although x has a binomial distribution with $n = 1000$ and $\pi = 1/200 = .005$, its distribution can be approximated by a Poisson distribution with $\lambda = n\pi = 1000/200 = 5$. Therefore, using Table III (p. 565), Proportion($x \geq 8$) = .065 + .036 + .018 + .008 + .003 + .001 = .131. Note: because the Table entries are rounded to 3 places, you would get a slightly different answer (of .135) if you worked the problem by first adding the proportions for $x \leq 7$, then subtracting from 1.

Proportion($5 \leq x \leq 10$) = .175 + .146 + .104 + .065 + .036 + .018 = .544.

61. (a) Let x = fracture strength. Then, Proportion($x < 90$) = .50 because the normal distribution is symmetric around its mean, which is $\mu = 90$.

Proportion($x < 95$) = Proportion($z < (95-90)/3.75$) = Proportion($z < 1.33$) = .9082 (from Table I).

Proportion($x \geq 95$) = 1 - Proportion($x < 95$) = 1 - .9082 = .0918.

(b) Proportion($85 \leq x \leq 95$) = Proportion($(85-90)/3.75 \leq z \leq (95-90)/3.75$) = Proportion($-1.33 \leq z \leq 1.33$) = .9082 - .0918 = .8164 (Table I).

Proportion($80 \leq x \leq 100$) = Proportion($(80-90)/3.75 \leq z \leq (100-90)/3.75$) = Proportion($-2.67 \leq z \leq 2.67$) = .9962 - .0038 = .9924 (Table I).

(c) Let x^* denote the value that is exceeded by 90% of the data. From Table I, the 10th percentile of the standard normal distribution is $z^* = -1.28$. Therefore, x^* must be 1.28 standard deviations below the mean of the x distribution. That is, $x^* = 90 - 1.28(3.75)$ = 85.20.

(d) The corresponding interval for the standard normal distribution is .99 = Proportion($-z^* \leq z \leq z^*$), which means that the left tail area is Proportion($z \leq z^*$) = .99 + .005 = .9950, so (from Table I) $z^* = 2.58$. Therefore, the corresponding two values for the x distribution are $90 \pm 2.58(3.75)$ = 80.325 and 99.675.

63. (a) Proportion(x > 1000) = Proportion(ln(x) > ln(1000)) = Proportion(ln(x) > 6.90776). Standardizing, this proportion equals Proportion(z > (6.90776-6.5)/.75) = Proportion(z > .54) = 1 - Proportion(z ≤ .54) = 1 - .7054 = .2946. Likewise, Proportion(x > 2000) = Proportion(z > (ln(2000)-6.5)/.75) = Proportion(z > 1.47) = 1 - Proportion(z ≤ 1.47) = 1-.9292 = .0708. Finally, Proportion(x > 3000) = Proportion(z > (ln(3000)-6.5)/.75) = Proportion(z > 2.01) = 1 - .9778 = .0222.

(b) Proportion(2500 ≤ x ≤ 5000) = Proportion((ln(2500)-6.5)/.75 ≤ z (ln(5000)-6.5)/.75) = Proportion(1.77 ≤ z ≤ 2.69) = Proportion(z ≤ 2.69) - Proportion(z ≤ 1.77) = .9964 - .9616 = .0348.

(c) Let x* denote the value that exceeds the fastest 10% of times. Then, .10 = Proportion(x < x*) = Proportion(ln(x) < ln(x*)) = Proportion(z < (ln(x*)-6.5)/.75). From Table I, the value of z exceeds the smallest 10% of the distribution is, approximately, z* = -1.28. Therefore, (ln(x*)-6.5)/.75 = -1.28, so ln(x*) = 6.5-1.28(.75) = 5.54. Exponentiating both sides, x* = exp(5.54) = 254.68.

(d) The density function appears below. The positive skewness is quite pronounced.

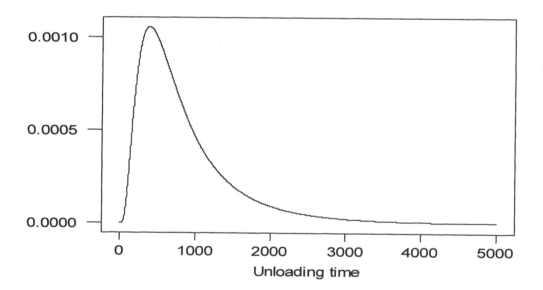

65. (a) Accommodating everyone who shows up means that x ≤ 100, so Proportion(x ≤ 100) = .05 + .10 + ... + .24 + .17 = .82.

(b) Proportion(x > 100) = 1 - Proportion(x ≤ 100) = 1 - .82 = .18.

19

(c) The first standby passenger will be able to fly as long as the number who show up is $x \le 99$, which leaves one free seat. This proportion is .65. The third person on the standby list will get a seat as long as $x \le 97$, where Proportion$(x \le 97) = .05 + .10 + .12 = .27$.

67.　(a) The area under the density must equal 1. To verify this, make the substitution $u = x^2/(2\theta^2)$,

which gives $du = xdx/\theta^2$, so $\int_0^{\infty} \frac{x}{\theta^2} e^{-x^2/(2\theta^2)} dx = \int_0^{\infty} e^{-u} du = -e^{-u} \Big]_0^{\infty} = 0 - (-e^0)$

$= 1$. For a finite upper limit $x = c$, the integral becomes: Proportion$(x \le c) =$

$\int_0^c \frac{x}{\theta^2} e^{-x^2/(2\theta^2)} dx = -e^{-u} \Big]_0^{c^2/(2\theta^2)} = 1 - e^{-c^2/(2\theta^2)}$.

(b) For $\theta = 100$, Proportion$(x \le 200) = 1 - e^{-200^2/(2(100)^2)} = 1 - e^{-2} = 1 - .135335 = .8647$.

Proportion$(x \ge 200) = 1 - $ Proportion$(x < 200) = 1 - .8647 = .1353$.

Proportion$(100 \le x \le 200) = $ Proportion$(x \le 200) - $ Proportion$(x \le 100) = .8647 - [1 - e^{-100^2/(2(100)^2)}] = .8647 - [1 - e^{-.5}] = .4712$.

69.　(a) The density curve is shown below. In essence, the curve is just an exponential density that has been shifted to the right by 0.5 units.

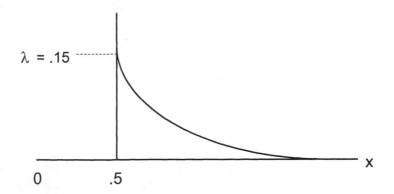

The total area under the density curve is $\int_{-\infty}^{\infty} f(x)dx = \int_{.5}^{\infty} .15e^{-.15(x-.5)} dx$. Using the

substitution y = x-.5, the last integral becomes $\int_{0}^{\infty} .15e^{-.15y} dy$, which equals 1 because it is

the area under an exponential density curve with $\lambda = .15$.

(b) Using the same substitution as in part (a), Proportion(x ≤ 5) = $\int_{5}^{5} .15e^{-.15(x-.5)} dx =$

$\int_{0}^{4.5} .15e^{-.15y} dy = 1 - e^{-.15(4.5)} = .491$. Similarly, Proportion(5 < x < 10) =

$\int_{5}^{10} .15e^{-.15(x-.5)} dx = \int_{4.5}^{9.5} .15e^{-.15y} dy = e^{-.15(4.5)} - e^{-.15(9.5)} = .269$.

(c) Let x^* denote the median of the headway times. Then, $.50 = \int_{.5}^{x^*} f(x)dx =$

$\int_{0}^{x^*-.5} .15e^{-.15y} dy = 1 - e^{-.15(x^*-.5)}$, so $e^{-.15(x^*-.5)} = 1 - .50 = .50$. Taking logarithms

of both sides we find $-.15(x^*-.5) = \ln(.50)$, so $x^* = .5 + \ln(.50)/(-.15) = 5.12$.

(d) Let x^* denote the 90th percentile of the headway times. Then, $.90 = \int_{.5}^{x^*} f(x)dx$.

Following the same steps as in part (c) we find $x^* = .5 + \ln(1-.90)/(-.15) = 15.85$.

71.   (a)   The cumulative proportion graph is shown below:

cumulative proportion

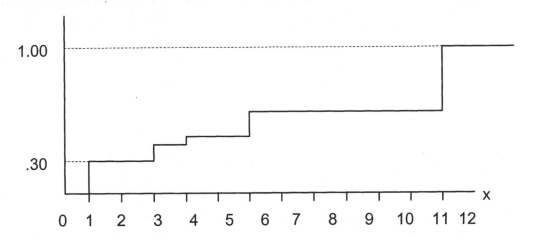

    (b)   x   =   1   3   4   6   12
            p(x) = .30   .10   .05   .15   .40

            For example, Proportion(x = 3) = Proportion(x ≤ 3) - Proportion(x < 3) =
            .40 - .30 = .10

    (c)   Proportion(3 ≤ x ≤ 6) = Proportion(x ≤ 6) - Proportion(x ≤ 3) = .60 - .30 = .30.

73.   Letting X = "bursting strength", we first find the proportion of all bottles having bursting
       strengths exceeding 300 psi.   Proportion(X > 300) = Proportion(z > (300-250)/30)) =
       Proportion (z > 1.67) = .0475 (from Table I). Then, Y = "the number of bottles in a carton
       of 12 with bursting strengths over 300 psi" is a binomial variable with n = 12 and π = .0475.
       So the proportion of all cartons containing at least one bottle with a bursting strength over
       300 psi is Proportion(Y ≥ 1) = 1 – Proportion(Y = 0) = 1 - $(1-p)^{12}$ = 1 – $(1- .0475)^{12}$ = .4423.

Chapter 2
Numerical Summary Measures

1.   (a) The sum of the $n = 11$ data points is 514.90, so $\bar{x} = 514.90/11 = 46.81$.

(b) The sample size ($n = 11$) is odd, so there will be a middle value. Sorting from smallest to largest: 4.4 16.4 22.2 30.0 33.1 36.6 40.4 66.7 73.7 81.5 109.9. The sixth value, 36.6 is the middle, or median, value. The mean differs from the median because the largest sample observations are much further from the median than are the smallest values.

(c) Deleting the smallest ($x = 4.4$) and largest ($x = 109.9$) values, the sum of the remaining 9 observations is 400.6. The trimmed mean $\bar{x}_{tr}$ is $400.6/9 = 44.51$. The trimming percentage is $100(1/11) \approx 9.1\%$. $\bar{x}_{tr}$ lies between the mean and median.

3.   (a) A stem-and leaf display of this data appears below:

```
Stem-and-leaf of C1          N  = 26
Leaf Unit = 1.0

    2     32 55
    4     33 49
    4     34
    8     35 6699
   13     36 34469
   13     37 03345
    8     38 9
    7     39 2347
    3     40 23
    1     41
    1     42 4
```

The display is reasonably symmetric, so the mean and median will be close.

(b) The sample mean is $\bar{x} = 9638/26 = 370.7$. The sample median is $\tilde{x} = (369+370)/2 = 369.50$.

(c) The largest value (currently 424) could be increased by any amount. Doing so will not change the fact that the middle two observations are 369 and 170, and hence, the median will not change. However, the value $x = 424$ can not be changed to a number less than 370 (a change of $424-370 = 54$) since that *will* lower the values(s) of the two middle observations.

(d) Expressed in minutes, the mean is (370.7 sec)/(60 sec) = 6.18 min; the median is 6.16 min.

5. The sum of the n = 10, observations is 120.10. so the mean is 12.01. The sorted data is: 6.5 7.9 9.2 10.0 10.7 12.0 12.5 14.5 14.9 21.9, so the middle two values are 10.7 and 12.0. Therefore, the median is (10.7+12.0)/2 = 11.35. Trimming off the largest and smallest values (i.e, using a 100(1/10)% = 10% trimming factor), the trimmed mean is (120.10-6.5-21.9)/8 = 11.46. Because the trimmed mean and median are close, either of these values (11.35 or 11.46) would be a good representative value for the data.

7. The sorted data are: 17, 29, 35, 48, 57, 79, 86, 92, 100+, 100+. The median is the average of the two middle observations: $\tilde{x}$ = (57+79)/2 = 68. On the other hand, the sample mean $\bar{x}$ can not be computed because the two largest values are not known (we only know that they are greater than 100). Several trimmed means can be computed: for example, by dropping the smallest and largest *two* observations (i.e., with 20% trimming), the trimmed mean is (35 +48+ +...+86+92)/8 = 397/8 = 66.2. Similarly, the 30% trimmed mean is 67.5.

9. (a) $\mu = \int_0^2 (x)(.5x)dx = .5\dfrac{x^3}{3}\Big]_0^2$ = 4/3. The mean μ does not equal 1 because the density curve is not symmetric around x = 1.

(b) Half the area under the density curve to the left (or right) of the median $\tilde{\mu}$, so, .50 =

$\int_0^{\tilde{\mu}} .5xdx = .5\dfrac{x^2}{2}\Big]_0^{\tilde{\mu}}$ = $(\tilde{\mu}^2)/4$, or, $\tilde{\mu}^2 = 4(.5) = 2$ and $\tilde{\mu} = 1.414$. $\mu < \tilde{\mu}$ because the density curve is negatively skewed.

(c) $\mu \pm \frac{1}{2} = \frac{4}{3} \pm \frac{1}{2} = \frac{5}{6}$ and $\frac{11}{6}$. The area under the curve between these two values is

$\int_{5/6}^{11/6} .5xdx = \dfrac{x^2}{4}\Big]_{5/6}^{11/6}$ = .615. Similarly, the proportion of the times that are within one-

half hour of $\tilde{\mu}$ is: $\int_{.914}^{1.914} .5xdx = \dfrac{x^2}{4}\Big]_{.914}^{1.914}$ = .707.

11. $\mu = \int_1^2 2x(1 - \dfrac{1}{x^2})dx = \int_1^2 2x\,dx - \int_1^2 \dfrac{2}{x}dx = x^2\Big]_1^2 - 2\ln(x)\Big]_1^2 = (2^2 - 1^2) - 2(\ln(2) - \ln(1))$

$= 3 - 2(.693147) = 1.61371$, or about 1.614. For the median, $\tilde{\mu}$, half of the area under the

density must lie to the left of $\tilde{\mu}$, so $.5 = \int_1^{\tilde{\mu}} 2(1 - \dfrac{1}{x^2})dx = 2x\Big]_1^{\tilde{\mu}} + \dfrac{2}{x}\Big]_1^{\tilde{\mu}} = 2(\tilde{\mu} - 1) + 2(1/\tilde{\mu}$

$- 1)$. Solving for $\tilde{\mu}$ results in a quadratic equation $2\tilde{\mu}^2 - 4.5\tilde{\mu} + 2 = 0$, whose roots are .61

and 1.64; only the root $\tilde{\mu} = 1.64$ is feasible (since it lies in the interval from 1 to 2). The

proportion of x values that lie within the mean and median is $\int_{1.614}^{1.64} 2(1 - \dfrac{1}{x^2})dx =$

$2x\Big]_{1.614}^{1.64} + \dfrac{2}{x}\Big]_{1.614}^{1.64} = .032.$

13. $\mu = \sum_{x=0}^{4} xp(x) = 0(.4) + 1(.1) + 2(.1) + 3(.1) + 4(.3) = 1.8$

15. (a) $\bar{x} = \dfrac{1}{n}\sum_i x_i = 577.9/5 = 115.58$. Deviations from the mean: $116.4 - 115.58 = .82$,

$115.9 - 115.58 = .32$, $114.6 - 115.58 = -.98$, $115.2 - 115.58 = -.38$, and $115.8 - 115.58 = .22$.

(b) $s^2 = [(.82)^2 + (.32)^2 + (-.98)^2 + (-.38)^2 + (.22)^2]/(5-1) = 1.928/4 = .482$, so $s = .694$.

(c) $\sum_i x_i^2 = 66,795.61$, so $s^2 = \dfrac{1}{n-1}\left[\sum_i x_i^2 - \dfrac{1}{n}\left(\sum_i x_i\right)^2\right] =$

$[66,795.61 - (577.9)^2/5]/4 = 1.928/4 = .482.$

17. (a) $\bar{x} = \dfrac{1}{n}\sum_i x_i = 14438/5 = 2887.6$. The sorted data is: 2781 2856 2888 2900 3013,

so the sample median is $\tilde{x} = 2888$.

(b) Subtracting a constant from each observation shifts the data, but does not change its
sample variance (Exercise 16(b)). For example, by subtracting 2700 from each
observation we get the values 81, 200, 313, 156, and 188, which are smaller (fewer
digits) and easier to work with. The sum of squares of this transformed data is 204210
and its sum is 938, so the computational formula for the variance gives
$s^2 = [204210 - (938)^2/5]/(5-1) = 7060.3.$

25

19.	Using the computational formula, $s^2 = \frac{1}{n-1}\left[\sum_i x_i^2 - \frac{1}{n}\left(\sum_i x_i\right)^2\right] =$

[3,587,566-(9638)²/26]/(26-1) = 593.3415, so s = 24.36. In general, the size of a typical deviation from the sample mean (370.7) is about 24.4. Some observations may deviate from 370.7 by a little more than this, some by less.

21.	(a) HC data: $\sum_i x_i^2 = 2618.42$ and $\sum_i x_i = 96.8$, so $s^2 = [2618.42 - (96.8)^2/4]/3 =$

91.953 and the sample standard deviation is s = 9.59. CO data: $\sum_i x_i^2 = 145645$ and

$\sum_i x_i = 735$, so $s^2 = [145645 - (735)^2/4]/3 = 3529.583$ and the sample standard

deviation is s = 59.41.

(b)	The mean of the HC data is 96.8/4 = 24.2; the mean of the CO data is 735/4 = 183.75. Therefore, the coefficient of variation of the HC data is 9.59/24.2 = .3963, or 39.63%. The coefficient of variation of the CO data is 59.41/183.75 = .3233, or 32.33%. Thus, even though the CO data has a larger standard deviation than does the HC data, it actually exhibits *less* variability (in percentage terms) around its average than does the HC data.

23.	(a) $\sigma = \sqrt{\lambda} = \sqrt{5} = 2.236$, so x values that lie between 5 - 2.236 = 2.764 and 5 + 2.236 = 7.236 are within one standard deviation from the mean. Because x is integer-valued, only the integers between 3 and 7 satisfy this requirement, i.e., Proportion(2.764 ≤ x ≤ 7.236) = Proportion(3 ≤ x ≤ 7) = p(3) + p(4) + p(5) +p(6) + p(7) = .140 + .175 + .175 + .146 + .104 = .740 (using Table III, p. 565, with λ = 5).

(b)	To exceed the mean by *more* than 2 standard deviations, x values must be greater than 5 + 2(2.236) = 9.472. The integer values of x than satisfy this requirement are x = 10 and greater. From Table III, Proportion(x > 9.472) = Proportion(x ≥ 10) = .018 +.008 + .003 +.001 = .030. Note: because table entries are rounded to 3 places, a slightly different answer results if you calculate the proportion as 1 - Proportion(x ≤ 9) = 1 - .966 = .032.

25.	$\sigma^2 = \Sigma(x-\mu)^2 p(x) = \Sigma(x^2 - 2\mu x + \mu^2)p(x) =$

$\Sigma x^2 p(x) - 2\mu\Sigma xp(x) + \mu^2\Sigma p(x) = \Sigma x^2 p(x) - 2\mu^2 + \mu^2 = \Sigma x^2 p(x) - \mu^2$.

For the mass function given in Exercise 24, $\sigma^2 = \Sigma x^2 p(x) - \mu^2 = (0)^2(.4) + (1)^2(.1)$
$+(2)^2(.1) + (3)^2(.1) +(4)^2(.3) - (1.8)^2 = 6.2 - 3.24 = 2.96$.

27. The proportion of x values between $\mu - 1.5\sigma$ and $\mu + 1.5\sigma$ is the same as the proportion of z values between -1.5 and +1.5: Proportion$(-1.5 \leq z \leq 1.5)$ = Proportion$(z \leq 1.5)$ - Proportion$(z \leq -1.5)$ = .9332-.0668 = .8664. The proportion of x value that exceed μ by more than 2.5 σ's equals Proportion$(z > 2.5)$ = 1 - Proportion$(z \leq 2.5)$ = 1 - .9938 = .0062.

29. The mean and variance of the lognormal distribution are $\mu_x = e^{\mu+\sigma^2/2}$ and $\sigma_x^2 = e^{2\mu+\sigma^2}(e^{\sigma^2} - 1)$, where μ and σ are the parameters of the normal distribution that describes $\ln(x)$. Therefore, set $900 = \mu_x$ and $725^2 = \sigma_x^2$ and solve the two equations in the two unknowns. Notice that $\mu_x^2 = e^{2\mu+\sigma^2}$, which can be substituted into the expression for σ_x^2: $725^2 = \mu_x^2((e^{\sigma^2} - 1) = 900^2(e^{\sigma^2} - 1)$. So, $(e^{\sigma^2} - 1) = (725/900)^2$, which yields $e^{\sigma^2} = 1 + (725/900)^2$. Taking logarithms of both sides, $\sigma^2 = \ln(1+((725/900)^2) = .50012$, so $\sigma = .7072$. Putting $\sigma = .7072$ into the equation for μ_x, we get $900 = e^{\mu+\sigma^2/2} = e^{\mu+(.7072)^2/2}$, so $\ln(900) = \mu + (.7072)^2/2$, which yields $\mu = 6.5523$. As a check on the calculations, when substituted into the original equations for mean and variance, the values $\mu = 6.5523$ and $\sigma = .7072$ do give $\mu_x = 900$ and $\sigma_x^2 = 725^2$.

31. $\bar{x} = 5049.16$ and $s = 351.45$, so:

interval rule %	actual %	empirical
$\bar{x} \pm 1(s) = 5049.16 \pm 1(351.45) = [4697.71, 5400.61]$	68/100 = 68%	68%
$\bar{x} \pm 2(s) = 5049.16 \pm 2(351.45) = [4346.26, 5752.06]$	96/100 = 96%	95%
$\bar{x} \pm 3(s) = 5049.16 \pm 3(351.45) = [3994.81, 6103.51]$	100/100 = 100%	99.7%

33. (a) The mean and median are nearly equal and the upper and lower quartiles are nearly equidistant from the median, so the data is approximately symmetrically distributed around the median (and mean).

   (b) The IQR = 138.25-132.95 = 5.3, so 1.5(IQR) = 7.95 and 3(IQR) = 15.9. Since the minimum value is 132.95 – 122.20 = 10.75 units away from the lower quartile and the maximum value is 147.70 - 138.25 = 9.45 above the upper quartile, then the values 122.20 and 147.70 both qualify as mild outliers.

35. (a) Lower half of the data set: 325 325 334 339 356 356 359 359 363 364 364 366 369, whose median, and therefore the lower quartile, is 359 (the 7th observation in the sorted list). The top half of the data is 370 373 373 374 375 389 392 393 394 397 402 403 424, whose median, and therefore the upper quartile is 392. So, the IQR = 392 - 359 = 33.

   (b) 1.5(IQR) = 1.5(33) = 49.5 and 3(IQR) = 3(33) = 99. Observations that are further than 49.5 below the lower quartile (i.e., 359-49.5 = 309.5 or less) or more than 49.5 units above the upper quartile (greater than 392+49.5 = 441.5) are classified as 'mild' outliers. 'Extreme' outliers would fall 99 or more units below the lower, or above the upper, quartile. Since the minimum and maximum observations in the data are 325 and 424, we conclude that there are no mild outliers in this data (and therefore, no 'extreme' outliers either).

(c) A boxplot (created by Minitab) of this data appears below. There is a slight positive skew to the data, but it is not far from being symmetric. The variation, however, seems large (the spread 424-325 = 99 is a large percentage of the median/typical value)

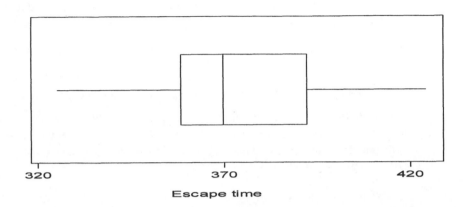

Escape time

(d) Not until the value x = 424 is lowered below the upper quartile value of 392 would there be any change in the value of the upper quartile. That is, the value x = 424 could not be decreased by more than 424-392 = 32 units.

37. (a) 1.5(IQR) = 1.5(216.8-196.0) = 31.2 and 3(IQR) = 3(216.8-196.0) = 62.4.
Mild outliers: observations below 196-31.2 = 164.6 or above 216.8+31.2 = 248.
Extreme outliers: observations below 196-62.4 = 133.6 or above 216.8+62.4 = 279.2.
Of the observations given, 125.8 is an extreme outlier and 250.2 is a mild outlier.

(b) A boxplot of this data appears below. There is a bit of positive skew to the data but, except for the two outliers identified in part (a), the variation in the data is relatively small.

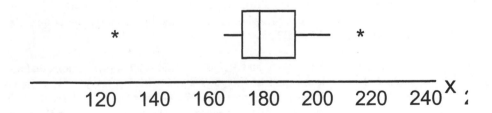

39.   (a) ED: median = .4 (the 14th value in the *sorted* list of data). The lower quartile (median of the lower half of the data, including the median, since n is odd) is (.1+.1)/2 = .1. The upper quartile is (2.7+2.8)/2 = 2.75. Therefore, IQR = 2.75 - .1 = 2.65.

Non-ED: median = (1.5+1.7)/2 = 1.6. The lower quartile (median of the lower 25 observations) is .3; the upper quartile (median of the upper half of the data) is 7.9. Therefore, IQR = 7.9 - .3 = 7.6.

(b) ED: mild outliers are less than .1 - 1.5(2.65) = -3.875 or greater than 2.75 + 1.5(2.65) = 6.725. Extreme outliers are less than .1 - 3(2.65) = -7.85 or greater than 2.75 + 3(2.65) = 10.7. So, the two largest observations (11.7, 21.0) are extreme outliers and the next two largest values (8.9, 9.2) are mild outliers. There are no outliers at the lower end of the data.

Non-ED: mild outliers are less than .3 - 1.5(7.6) = -11.1 or greater than 7.9 + 1.5(7.6) = 19.3. Note that there are no mild outliers in the data, hence there can not be any extreme 19.4. outliers either.

(c) A comparative boxplot appears below. The outliers in the ED data are clearly visible. There is noticeable positive skewness in both samples; the Non-Ed data has more variability then the Ed data; the typical values of the ED data tend to be smaller than those for the Non-ED data.

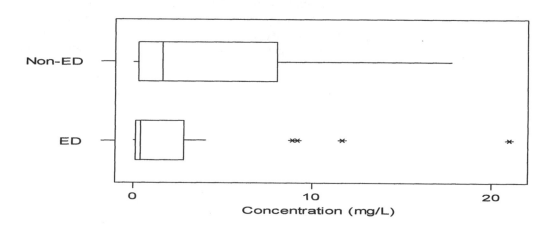

41.   Outliers occur in the 6 a.m. data. The distributions at the other times are fairly symmetric. Variability and the 'typical' values in the data increase a little at the 12 noon and 2 p.m. times.

43   (a) Let $x = SO_2$ concentration. Because $\ln(x)$ has a normal distribution, we can simply compute the quartiles of $\ln(x)$, then exponentiate to find the quartiles of x. The lower and upper quartiles of a normal distribution are (cf text, p.80): $\mu \pm .675\sigma = 1.9 \pm .675(.9) = 1.2925$ and 2.5075. Therefore, the lower and upper quartiles of x are $e^{1.2925} = 3.642$ and $e^{2.5075} = 12.274$.

(b) The 95th percentile of the normal distribution $\ln(x)$ is $\mu + 1.645\sigma = 1.9 + 1.645(.9) = 3.3805$, so the 95th percentile of x's distribution is $e^{3.3805} = 29.386$.

(c) For $\mu = 2.0$, the new quartiles become $e^{2-.675(.9)} = 4.0259$ and $e^{2+.675(.9)} = 13.561$, compared to the previous lower and upper quartiles of 3.642 and 12.274. The increases, $4.0259 - 3.642 = .38$ and $13.561 - 12.274 = 1.29$, are not the same.

45.   The normal quantiles are easy to generate using Minitab or Excel. For example, in Minitab, typing the following commands will generate the normal quantiles shown on page 89 of the text:

```
MTB> set c2
MTB> 1:20
MTB> let c3 = (c2-.5)/20
MTB> invcdf c3 c4;
SUBC> norm 0 1.
```

(Note: you don't have to type 'END' to end the input of data into column C2; typing any valid Minitab command will automatically end data input and then will execute the command)

Although it isn't necessary in this exercise, remember to sort (from smallest to largest) the data before plotting it versus the normal quantiles. The quantile plot for this data is shown below. The pattern is obviously nonlinear, so a normal distribution is implausible for this data. The apparent break that appears in the data in the right side of the graph is indicative of data that contains outliers.

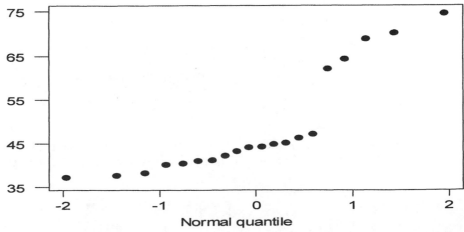

47.   The quantile plot (created using the same technique as in Exercise 45) is shown below. The plot is nonlinear, primarily because the largest value in the data may be an outlier or because the actual distribution of this data is positively skewed. In any case, a normal distribution is not the best fit to this data.

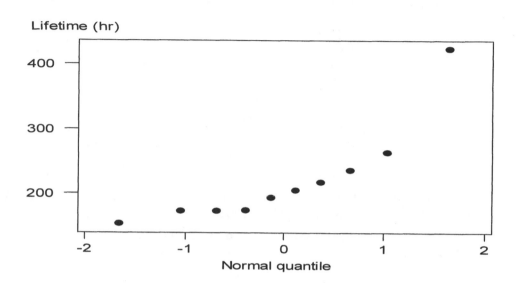

49.   The Weibull quantile plot data can be created in Minitab or Excel. For example, the following Minitab commands will generate the natural logarithms of the x data (assumed to be stored in column C1), the percentiles p_i, and the Weibull plotting points $\ln(-\ln(1-p_i))$:

```
MTB > let c10 = loge(c1)
MTB > set c2
DATA> 1:11
DATA> let c3 = (c2-.5)/11
MTB > let c4 = loge(-loge(1-c3))
MTB > plot c10*c4
```

The Weibull plot created from this data appears below. The points do not show a marked deviation from linearity, so a Weibull distribution would be plausible for this data.

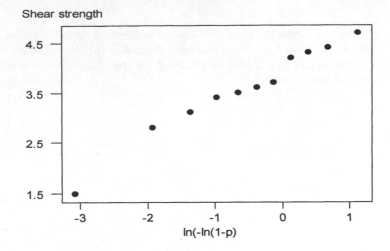

Shear strength

ln(-ln(1-p))

51. Let η_p denote the pth quantile of an exponential distribution with parameter λ. Then, the area to the *right* of η_p is 1-p. Recall from Exercise 23 of Chapter 1, that the right tail area (i.e., the area past x = c) for an exponential distribution is simply $e^{-\lambda c}$. Therefore, $e^{-\eta_p \lambda}$ = 1-p. Taking logarithms of both sides, $-\eta_p \lambda = \ln(1-p)$, so $\eta_p = \lambda(-\ln(1-p))$. That is, the quantiles η_p are linearly related to the quantitites $-\ln(1-p)$, so a plot of the sample quantiles $x_{(i)}$ versus $-\ln(1-p_i)$ is a straight line.

In this exercise, n = 16, so the values of p_i equal (i-.5)/16 for i = 1,2,...16. Use Minitab or Excel to compute the plotting values $-\ln(1-p_i)$. The quantile plot for this data appears below. Because the plot exhibits curvature, an exponential distribution would not be appropriate for this data.

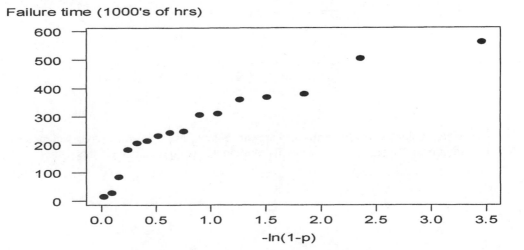

Failure time (1000's of hrs)

-ln(1-p)

53. (a)

	mean	median	standard deviation
PTSD:	32.92	37.00	9.93
Healthy:	52.23	51.00	14.86

(b)

	trimmed mean	
PTSD:	33.82	One observation is removed from each
Healthy:	53.09	end, so the trimming % is 1/13, or 7.7%.

Trimming removes the effect of any extreme observations in the data.

(c)

> Note: The definition of lower and upper quartile used in this text is slightly different than the one used by some other authors (and software packages). Technically speaking, the median of the lower half of the data is not really the first quartile, although it is generally *very close*. Instead, the medians of the lower and upper halves of the data are often called the **lower** and **upper hinges**. Our boxplots use the lower and upper hinges to define the spread of the middle 50% of the data, but other authors sometimes use the *actual* quartiles for this purpose. The difference is usually very slight, usually unnoticeable, but not always.

Using the text's definition, the *medians* of the two halves of each data set and the IQR are:

	Lower half	Upper half	IQR
PTSD:	28	39	39-28 = 11
Healthy:	41	66	66-41 = 25

The smallest observation of 10 in the PTSD data is $28 - 10 = 18$ units from the lower quartile (hinge) which exceeds 1.5(IQR) = 16.5 units and would make 10 a mild outlier. There are no other potential outliers.

Using Minitab's definition, the upper and lower quartiles and IQR are:

	Lower quartile	Upper quartile	IQR
PTSD:	26.5	39.0	39.0-26.5 = 12.5
Healthy:	40.5	66.5	66.5-40.5 = 26.0

With this definition, the observation 10 in the PTSD data is 26.5-10 = 16.5 units from the lower quartile, which does not exceed 1.5(IQR) = 18.75, so 10 would not be considered an outlier in this case. There are no outliers of any kind indicated using the Minitab definition.

(d)   Comparative boxplots produced with Minitab are shown below. Note that the middle 50% (i.e., the shaded box) of the Healthy data lies above the middle 50% of the PTSD data, so there is only a small amount of overlap between the values in these two samples. Furthermore, the PTSD exhibits smaller variation than the Healthy data.

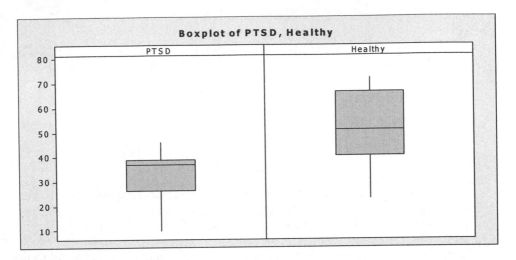

(e) The Heathly data is not very skewed, but the PTSD data is highly skewed (its median is very near the upper quartile). Therefore, statistical methods that presuppose the data comes from a normal distribution would not be recommended for this data.

55.

Flow rate	Median	Lower quartile	Upper quartile	IQR	1.5(IQR)	(IQR)
125	3.1	2.7	3.8	1.1	1.65	3.3
160	4.4	4.2	4.9	.7	1.05	2.1
200	3.8	3.4	4.6	1.2	1.80	3.6

There are no outliers in the three data sets. However, as the comparative boxplot below shows, the three data sets differ with respect to their central values (the medians are different) and the data for flow rate 160 is somewhat less variable than the other data sets. Flow rates 125 and 200 also exhibit a small degree of positive skewness.

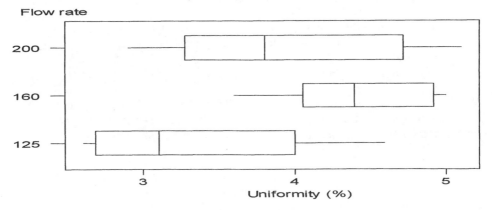

57.   Using the short-cut formula for s^2,

$$s^2 = \frac{1}{27-1}\left[24,657,511 - \frac{1}{27}(20,179)^2\right] = 368,320.1652 \,.$$ Taking square roots gives
$s = 606.894$. The mean is $\bar{x} = 20,179/27 = 747.3704$.

Therefore, the maximum that could be awarded is 2 standard deviations above the mean, or,

$\bar{x} + 2s = 747.3704 + 2(606.894) = 1961.158$. Since the data is given in units of $1000, this result can be expressed in dollars as $1,961,158.

59.   (a)   Percentage within 2 standard deviations is at least $100(1-1/2^2) = 75\%$.
               Percentage within 3 standard deviations is at least $100(1-1/3^2) = 89\%$.
               Percentage within 5 standard deviations is at least $100(1-1/5^2) = 96\%$.
               Percentage within 10 standard deviations is at least $100(1-1/10^2) = 99\%$.

        (b)   The percentage that are within 2 standard deviations is *at least* 75%, so the percentage that are further than 2 standard deviations from the mean is *at most* 25%. Similarly, *at most* 11% (i.e., 100% - 89%) of the values are further than 3 standard deviations from the mean.

        (c)   The values .995 and 1.005 lie at a distance of $(.995-1.000)/.0025 = -2$ standard deviations and $(1.005-1.000)/.0025 = +2$ standard deviations from the mean. Therefore, values that lie between .995 and 1.005 are ones that are within 2 standard deviations of the mean. Chebyshev's inequality indicates that *at least* 75% of the widths will lie in this interval; i.e., at least 75% of the slot widths will conform to specifications.

        (d)   The percentage of the observations *outside* the interval from .995 to 1.005 will be *at most* 100%-75% = 25%. Chebyshev's inequality makes no assumptions about the shape of the distribution of the values. If the distribution is symmetric, then it would be legitimate to say that about half, or 12.5%, of the values lie above 1.005 (and the other 12.5 lie below .995). However, if it is not known whether or not the distribution is symmetric (as is the case in this exercise), then it is possible that all 25% of the values could lie above 1.005 (or below .995). So, making no assumptions about the distribution, the best we can say is that the percentage of values that exceed 1.005 is at most 25%.

61. The transformation of the x_i's into the z_i's is analogous to 'standardizing' a normal distribution. The purpose of standardizing is to reduce a distribution (or, in this exercise, a set of data) to one that has a mean of 0 and a standard deviation of 1. To show that this transformation achieves this goal, note that:

$$\sum z_i = \sum \tfrac{1}{s}(x_i - \bar{x}) = \tfrac{1}{s}\sum(x_i - \bar{x}) = \tfrac{1}{s}(0) = 0, \text{ so, dividing this sum by n, } \bar{z} = 0.$$

Next, $\tfrac{1}{n-1}\sum(z_i - \bar{z})^2 = \tfrac{1}{n-1}\left(\tfrac{1}{s^2}\right)\sum(x_i - \bar{x})^2 = \left(\tfrac{1}{s^2}\right)s^2 = 1.$

63. (a) The median is the same (371) in each plot and all three data sets are very symmetric. In addition, all three have the same minimum value (350) and same maximum value (392). Moreover, all three data sets have the same lower (364) and upper quartiles (378). So, all three boxplots will be *identical*.

(b) A comparative dotplot is shown below. These graphs show that there are differences in the variability of the three data sets. They also show differences in the way the values are distributed in the three data sets.

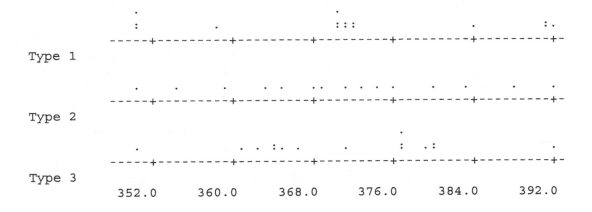

(c) The boxplot in (a) is not capable of detecting the differences among the data sets. The primary reason is that boxplots give up some detail in describing data because they use only 5 summary numbers for comparing data sets.

Recall that our text's definition of quartiles differs slightly from the one used by other authors (and by Minitab). See the comments about this in the answer to problem 53. A comparative boxplot based on the actual quartiles (as computed by Minitab) is shown below. The graph shows substantially the same type of information as those described in (a) except the graphs based on quartiles are able to detect the slight differences in variation between the three data sets.

Type of wire

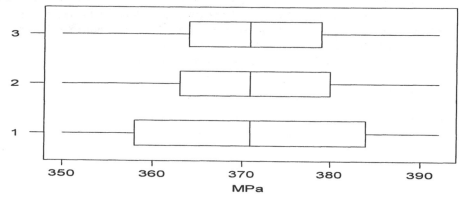

65. (a) Let y denote the capacitance of a capacitor. Capacitors will conform to specification when y is in the interval from 95 nf to 105 nf. Therefore, Proportion($95 \leq y \leq 105$) = Proportion((95-98)/$2 \leq z \leq (105$-98)/2) = Proportion($-1.5 \leq z \leq 3.5$). Using Table I, this proportion is equivalent to Proportion($z \leq 3.5$) - Proportion($z \leq -1.5$) = .9998 - .0068 = .9930, or, about 93.3%.

(b) The number of capacitors in a btach of 20 that conform to specifications will have a binomial distribution with n = 20 and π = .9330. Therefore, the proportion of batches containing at least 19 conforming capacitors is Proportion($x \geq 20$) = Proportion($x = 19$) + Proportion($x = 20$). Using the formula for the binomial mass function (page 50 of the text):

$$\frac{20!}{19!1!}(.9930)^{19}(1-.9930)^{1} + \frac{20!}{20!0!}(.9930)^{20}(1-.9930)^{0} = .9914$$

67. Assuming that the histogram is unimodal, then there is evidence of positive skewness in the data since the median lies to the left of the mean (for a symmetric distribution, the mean and median would coincide). For more evidence of skewness, compare the distances of the 5th and 95th percentiles from the median:

median - 5th percentile = 500 - 400 = 100 while 95th percentile -median = 720 - 500 = 220. Thus, the largest 5% of the values (above the 95th percentile) are further from the median than are the lowest 5%.

The same skewness is evident when comparing the 10th and 90th percentiles to the median: median - 10th percentile = 500 - 430 = 70 while 90th percentile -median = 640 - 500 = 140. Finally, note that the largest value (925) is much further from the median (925-500 = 425) than is the smallest value (500 - 220 = 280), again an indication of positive skewness.

69. Comparative boxplots of Summer and Winter relative humidity are shown below. From these plots it is apparent that the humidity tends to be higher during Winter, although the variation in humidity is about the same in both seasons. The second observation in the Winter data (41.83) is an outlier and should be examined to see if it is a valid observation.

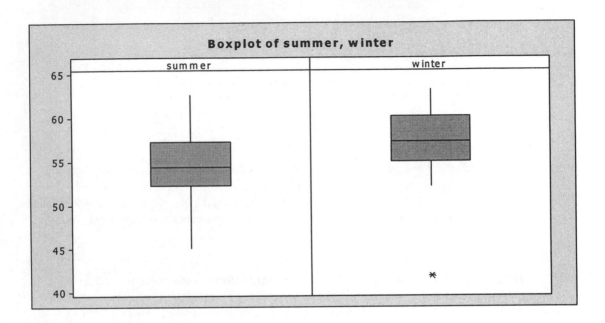

Chapter 3
Bivariate and Multivariate Data and Distributions

1.   (a) The following stem-and-leaf displays were created by Minitab:

```
Stem-and-leaf of temp      N  = 24
Leaf Unit = 1.0

    1     17 0
    3     17 23
    6     17 445
    8     17 67
    8     17
   (7)    18 0000011
    9     18 2222
    5     18 445
    2     18 6
    1     18 8
```

Each stem is repeated 5 times: the first stem of 17 has leaves of 0 and 1, the next stem has leaves 2 and 3, and so forth. 180 appears to be a typical value for this data. The distribution is reasonably symmetric in appearance and somewhat bell-shaped. The variation in the data is fairly small since the rang of the values (188-170 = 18) is fairly small compared to the typical value of 180.

```
Stem-and-leaf of ratio     N  = 24
Leaf Unit = 0.10

    3      0 889
    7      1 0000
    8      1 3
   12      1 4444
   12      1 66
   10      1 8889
    6      2 11
    4      2
    4      2 5
    3      2 6
    2      2
    2      3 00
```

39

For the 'ratio' data, a typical value is around 1.6 and the distribution appears positively skewed. The hundredths digit has been truncated in the display. The variation in the data is large since the range of the data (3.08 - .84 = 2.24) is very large compared to the typical value of 1.6. The two largest values could be outliers and should be checked using the methods of Section 2.3 (page 83 of the text).

(b) The efficiency ratio is *not* uniquely determined by temperature since there are several instances in the data of equal temperatures associated with different efficiency ratios. For example, the five observations with temperatures of 180 each have different efficiency ratios.

(c) A scatter plot of the data appears below. The points exhibit quite a bit of variation and do not appear to fall close to any straight line or simple curve.

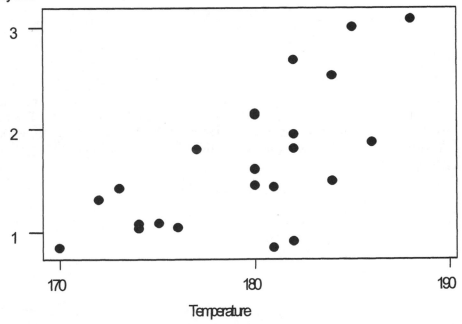

3. A scatter plot of the data appears below. The points fall very close to a straight line with an intercept of approximately 0 and a slope of about 1. This suggests that the two methods are producing substantially the same concentration measurements.

Concentration (sensor method)

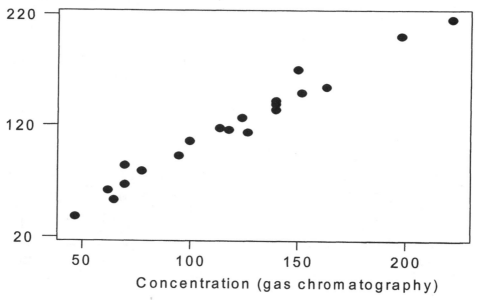

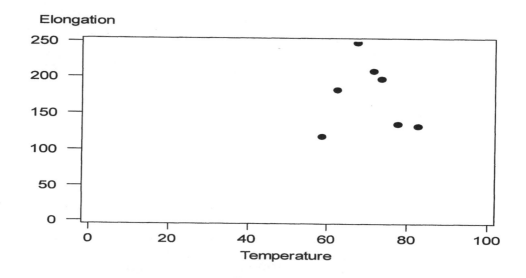

5.   (a) The scatter plot with axes intersecting at (0,0) is shown below.

(b) The scatter plot with axes intersecting at (55,100) appears below. The plot in (b) makes it somewhat easier to see the nature of the relationship between the two variables.

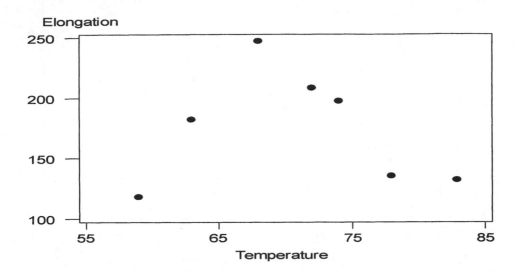

(c) A parabola appears to provide a good fit to both graphs.

7.   The scatter plot of this data is shown below. The plot indicates that there is a very strong negative relationship between the two variables; i.e., the eddy current response decreases in a linear fashion as the oxide-layer thickness is increased.

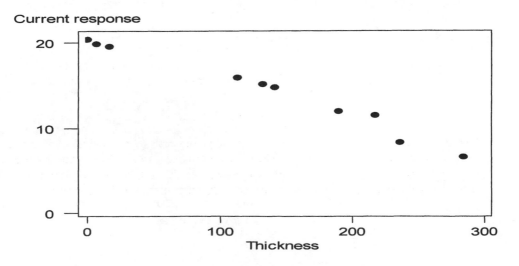

9. (a) Positive - As temperature increases, people make more use of air conditioners and fans, which will increase utility costs.

    (b) Negative - Higher interest rates makes loans more difficult to obtain and to pay off, which generally reduces the number of people seeking loans.

    (c) Positive - Married couples tend to have similar educational backgrounds and educational level is positively correlated with income level.

    (d) Over the entire range of speeds, there is little correlation: at very slow speeds, an increase in speed will improve fuel efficiency; at high speeds, additional increases in speed tend to reduce fuel efficiency.

    (e) Negative - Gasoline costs decrease as fuel efficiency increases.

    (f) Distance and GPA should be unrelated, so there should be little or no correlation.

11. (a) $SS_{xy} = 5530.92 - (1950)(47.92)/18 = 339.586667$, $SS_{xx} = 251{,}970 - (1950)^2/18 = 40{,}720$, and $SS_{yy} = 130.6074 - (47.92)^2/18 = 3.033711$, so

$$r = \frac{339.586667}{\sqrt{40720}\sqrt{3.033711}} = .9662$$

There is a very strong positive correlation between the two variables.

    (b) Because the association between the variables is positive, the specimen with the larger shear force will tend to have a larger percent dry fiber weight.

    (c) Changing the units of measurement on either (or both) variables will have no effect *on the calculated value of r*, because any change in units will affect both the numerator and denominator of r by exactly the same multiplicative constant.

13. Most people acquire a license as soon as they become eligible. If for, example, the minimum age for obtaining a license is 16, then the time since acquiring a license, y, is usually related to age by the equation $y \approx x- 16$, which is the equation of a straight line. In other words, the majority of people in a sample will have y values that closely follow the line $y = x-16$.

15.   Let d_0 denote the (fixed) length of the stretch of highway. Then, d_0 = distance = (rate)(time) = xy. Dividing both sides by x, gives the equation $y = d_0/x$ which means the relationship between x and y is curvilinear (in particular, the curve is a hyperbola). However, for values of x that are fairly close top one another, sections of this hyperbola can be approximated very well by a straight line with a negative slope (to see this, draw a picture of the function d_0/x for a particular value of d_0). This means that r should be closer to -.9 than to any of the other choices.

17.   (a) $SS_{xx} = 37695 - (561)^2/9 = 2726$, $SS_{yy} = 40223 - (589)^2/9 = 1676.222$, and
        $SS_{xy} = 38281- (561)(589)/9 = 1566.666$, so

$$r = \frac{1566.667}{\sqrt{2726}\sqrt{1676.222}} = .733.$$

   (b) $\bar{x}_1 = (70+72+94)/3 = 78.667$, $\bar{y}_1 = (60+83+85)/3 = 76.$
        $\bar{x}_2 = (80+60+55)/3 = 65$, $\bar{y}_2 = (72+74+58)/3 = 68.$
        $\bar{x}_3 = (45+50+35)/3 = 43.333$, $\bar{y}_3 = (63+40+54)/3 = 52.333.$

   $SS_{\overline{xx}} = [(78.667)^2+(65)^2+(43.333)^2 - (78.667+65+43.333)^2/3] = 634.913,$
   $SS_{\overline{yy}} = [(76)^2+(68)^2+(52.333)^2-(76+68+52.333)^2/3] = 289.923,$
   $SS_{\overline{xy}} = [(78.667)(76)+(65)(68)+(43.333)(52.333)-(187)(196.333)/3] = 428.348,$ so

$$r = \frac{428.348}{\sqrt{634.913}\sqrt{289.923}} = .9984.$$

   (d) The correlation among the averages is noticeably higher than the correlation among the raw scores, so these points fall much closer to a straight line than do the unaveraged scores. The reason for this is that averaging tends to reduce the variation in data, making it more likely that the averages will fall close to a straight line than the more variable raw data.

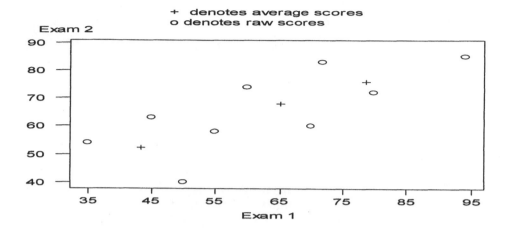

+ denotes average scores
o denotes raw scores

19. For this data: $n = 4$, $\sum x_i = 200$, $\sum y_i = 5.37$, $\sum x_i y_i = 333$, $\sum x_i^2 = 12,000$, and $\sum y_i^2 = 9.3501$. So, $SS_{xx} = 12,000 - (200)^2/4 = 2000$, $SS_{yy} = 9.3501 - (5.37)^2/4 = 2.140875$, and $SS_{xy} = 333 - (200)(5.37)/4 = 64.5$.

Therefore, $b = SS_{xy}/SS_{xx} = 64.5/2000 = .03225$ and $a = (5.37/4) - (.03225)(200/4) = -.27000$;
$SSResid = SSTo - bSS_{sy} = SS_{yy} - bSS_{xy} = 2.140875 - (.03225)(64.5) = .060750$;
$r^2 = 1 - SSResid/SSTo = 1 - .060750/2.140875 = .972$. This is a very high value of r^2, which confirms the authors' claim that there is a strong linear relationship between the two variables.

21. (a) The following stem-and-leaf display shows that: a typical vale for this data is a number in the low 40s, there is some positive skew in the data, there are some potential outliers (79.5 and 80.0), and there is a reasonable large amount of variation in the data (e.g., the spread 80.0 - 29.8 = 50.2 is large compared the typical values in the low 40s.

```
         Stem-and-leaf of MoE      N  = 27
         Leaf Unit = 1.0
             1      2 9
             3      3 33
            13      3 5566677889
            (4)     4 1223
            10      4 56689
             5      5 1
             4      5
             4      6 2
             3      6 9
             2      7
             2      7 9
             1      8 0
```

45

(b) No, the strength values are not uniquely determined by the MoE values. For example, note that the two pairs of observations having strength values of 42.8 have different MoE values.

(c) The least squares line is $\hat{y} = 3.2925 + .10748x$. For a beam whose modulus of elasticity is $x = 40$, the predicted strength would be $\hat{y} = 3.2925 + .10748(40) = 7.59$. The value $x = 100$ is far beyond the range of the x values in the data, so it would be dangerous (i.e., potentially misleading) to extrapolate the linear relationship that far.

(d) From the printout, SSResid = 18.736, SSTo = 71.605, and the coefficient of determination is $r^2 = .738$ (or, 73.8%). The r^2 value is large, which suggests that the linear relationship is a useful approximation to the true relationship between these two variables.

(e) There is no obvious pattern in the residuals that might suggest anything other than an approximately linear relationship between the variables.

23.   (a) From the 'Parameter Estimate' column of the printout, the least squares line is $\hat{y} = 3.620906 - 0.014711x$, where x = fracture strength. Substituting the value $x = 50$ into the equation gives a predicted attenuation of $\hat{y} = 3.620906 - 0.014711(50) = 2.8854$, or, about 2.89.

(b) From the 'Sum of Squares' column of the printout, SSResid = .26246 and SSTo = 2.55714. The r^2 value is .8974, or 89.74%. s_ε is called the 'Root MSE' (where MSE stands for 'mean square error') in the printout, so $s_\varepsilon = .14789$. The high vale of r^2 and the small value of s_ε (compared to the typical size of the y data values) indicate that the least squares line effectively summarizes the relationship between the variables.

(c) The plot of the residuals versus x (below) shows no discernible patterns, suggesting that no modification to the straight-line model is needed.

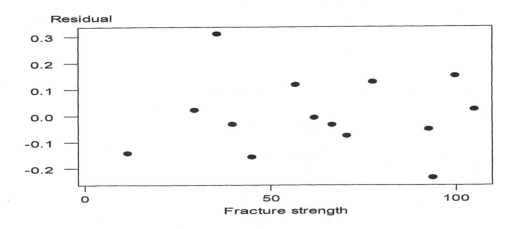

25. (a) The scatter plot shows a linear trend and the variation of the points appears to be reasonably constant for all values of the x (temperature) variable, so a simple linear regression model is appropriate for this data.

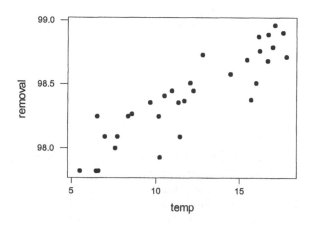

(b)

```
The regression equation is
removal = 97.5 + 0.0757 temp

Predictor          Coef         StDev             T          P
Constant        97.4986        0.0889       1096.17      0.000
temp           0.075691       0.007046        10.74      0.000

S = 0.1552      R-Sq = 79.4%      R-Sq(adj) = 78.7%

Analysis of Variance

Source             DF            SS            MS          F          P
Regression          1        2.7786        2.7786     115.40      0.000
Residual Error     30        0.7224        0.0241
Total              31        3.5010
```

When the temperature (x) is 10.50 °C, the predicted removal efficiency is:

$$\hat{y} = 97.4986 + 0.0757(10.50) = 98.2935, \text{ or, about } 98.29\%.$$

For observation 21, where the temperature is 10.50 °C, the actual y value is 98.41, so the residual for observation 21 is:

$$y_{21} - \hat{y}_{21} = 98.41 - 98.29 = 0.12$$

47

(c)   The typical deviation of points from the regression line is s_e, which is computed as follows:

$$s_e{}^2 = \frac{SS\,Re\,sid}{n-2} = \frac{.7224}{32-2} = .02408, \quad \text{so } s_e = \sqrt{.02408} = .1552.$$

(d)   The proportion of the variation in removal efficiency that can be attributed to the relationship with temperature is the coefficient of determination, r^2. For this data,

$$r^2 = 1 - \frac{SS\,Re\,sid}{SSTo} = 1 - \frac{.7224}{3.5010} = 1 - .20634 = .79366. \quad \text{Expressed as a percentage,}$$

about 79.4 % of the variation in removal efficiency can be explained by relationship between removal efficiency and temperature.

(e)   The point (6.53, 96.55) is an outlier that lies far below the rest of the data (whose y values are all between 97 and 98.96. Adding this point to the data will have a large influence on the regression (pulling it downward towards the outlier), which will cause the resulting estimated line to provide a worse fit the *rest* of the data. Therefore, the s_e value should become larger (in fact, it becomes .2911) and the r^2 should become smaller (in fact, it becomes 61.6%).

27.   (a) The plot is shown below. The plot suggests that a straight-line relationship log(time) $\approx$ a + b(log(edges) provides a good fit to the data. Letting y denote the recognition time and leting x denote the number of edges on a part, exponentiating both sides of this equation yields the approximate equation y $\approx$ kx^b (where k = 10^a, since logarithms base 10 were used). This is a 'power relationship' between x and y.

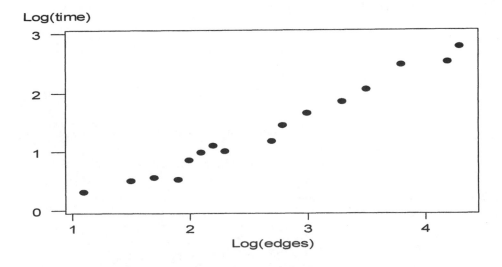

(b) The following printout (from Minitab) shows a least squares fit for this data. The large value of r^2 (97.5%) and the small value of s_ε (.1241, which is small compared to the typical y values) indicate that the least squares line provides an excellent fit to this data.

```
The regression equation is
logtime = - 0.760 + 0.798 loedg

Predictor        Coef        StDev          T         P
Constant      -0.76011      0.09330      -8.15     0.000
loedg          0.79839      0.03320      24.05     0.000

S = 0.1241       R-Sq = 97.6%      R-Sq(adj) = 97.5%

Analysis of Variance

Source        DF         SS          MS         F         P
Regression     1      8.9112      8.9112    578.23    0.000
Error         14      0.2158      0.0154
Total         15      9.1270
```

(c) Substituting the values of a = -.76011 an b = .79839 from the printout in (b) into the power relationship in part (a), the nonlinear relationship between y and x can be described by the equation $y \approx .17374x^{.79839}$ (where $.17374 = 10^{-.76011}$).

29. (a) The following plot of ln(y) versus x shows a fairly strong linear relationship. Following the lot is a Minitab printout of the least squares fit for this data. The large value of r^2 (87%) confirms the conclusions drawn from the scatterplot. Using the coefficients from the printout, the approximate relationship between ln(y) and x is: ln(y) $\approx$ -.5600 + .2382x. Exponentiating both sides gives the approximate equation $y \approx .5712e^{.2382x}$. Substituting the value x = 20, the predicted amount of soil erosion is $y \approx .5712 e^{.2382(20)}$ = 66.95, or, about 67.0 kg/day.

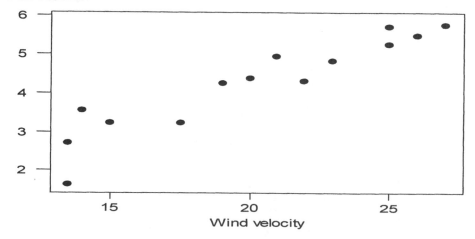

```
The regression equation is
logy = - 0.560 + 0.238 velocity

Predictor          Coef          StDev              T          P
Constant        -0.5600         0.5485          -1.02      0.327
velocity        0.23820        0.02658           8.96      0.000

S = 0.4616        R-Sq = 87.0%      R-Sq(adj) = 85.9%

Analysis of Variance

Source         DF           SS            MS          F          P
Regression      1       17.111        17.111      80.32      0.000
Error          12        2.556         0.213
Total          13       19.668
```

(b) A plot of the residuals (below) from the least squares fit in (a) shows no patterns that might suggest the choice of transformation of y is inappropriate.

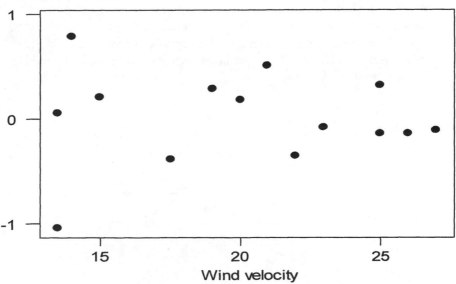

31. Letting x = Spectral Index and y = ln(L_{178}): n = 12, $\sum x_i = 9.72$, $\sum y_i = 313.1$, $\sum x_i y_i =$
255.107, $\sum x_i^2 = 8.0976$, and $\sum y_i^2 = 8182.81$. So, $SS_{xx} = 8.0976 - (9.72)^2/12 = .2244$,
$SS_{yy} = 8182.81 - (313.1)^2/12 = 13.509167$, and $SS_{xy} = 255.107 - (9.72)(313.1)/12 =$
1.4960. Therefore, the least squares coefficients are b = SS_{xy}/SS_{xx} = 1.4960/.2244 = 6.6667
and a = (313.1/12) - 6.6667(9.72/12) = 20.6917. The least squares line in then ln(L_{178}) ≈
20.6917 + 6.6667x. Exponentiating both sides gives L_{178} ≈ 968,927,163.1$e^{6.6667x}$.
Substituting x = .75 gives a predicted luminosity value of L_{178} ≈ 968,927,163.1$e^{6.6667(.75)}$ =
1.438051363 × 10^{11}.

33. (a) From the 'Coef' column in the printout, the least squares equation is $\hat{y}$ = -251.6 +
1000.1x -135.44x^2. For a water supply value of x = 3.0, the predicted wheat yield
would be $\hat{y}$ = -251.6 + 1000.1(3.0) -135.44$(3.0)^2$ = 1529.74.

     (b) From the printout, SSResid = 202,227 and SSTo = 1,372,436. These values yield an r^2
value of 1 - SSResid/SSTo = 1 - 202,227/1,372,436 = .8526, or, about 85.3%. The
large value of r^2 suggests that the quadratic fit is good.

35. (a) The least squares line is $\hat{y}$ = 39.3467 + 1.0958x_1 + 6.7958x_2 + 3.0625x_3 + 1.8542x_4.
Substituting the values x_1 = 35, x_2 = 75, x_3 = 200, and x_4 = 20: $\hat{y}$ = 39.3467 +
1.0958(35) + 6.7958(75) + 3.0625(200) + 1.8542(20) = 1236.97.

     (b) r^2 = 1 - SSResid/Ssto = 1 - 6557.7/21005.9 = .688, or about 68.8%.

37. (a) Substituting the values x_1 = 5 and x_2 = 35 into the least squares equation,
$\hat{y}$ = -151.36 -16.22(5) + 13.48(35) + .094$(5)^2$ - .253$(35)^2$ + .492(5)(35) = 17.87

     (b) The point x_1 = 30 and x_2 = 35 is not close to the majority of the data points (see the
graph below), so the prediction from the least squares model will not as good in this region.

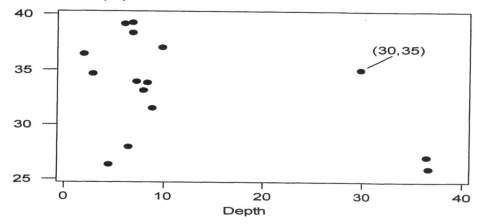

(c) $SSTo = \sum y_i^2 - (\sum y_i)^2/n = 7855.37 - (295.5)^2/14 = 1618.21$. $SSResid = \sum (y_i - \hat{y})^2 = (-.83487)^2 + (1.5755)^2 + \ldots + (4.8771)^2 = 394.64$. Therefore, $r^2 = 1 - SSResid/SSTo = 1 - 394.64/1618.21 = .756$, or about 76%.

(d) Yes, the $SSResid = 394.64$ from the model with all 5 variables (x_1, x_2, x_1^2, x_2^2, x_1x_2) is considerably lower than the $SSResid = 894.81$ from the two-variable model (x_1, x_2). Relative to the $SSTo$, this means that the 5-variable models "explains" much more of the variability in the y data, so it appears that at least one of the second-order terms provides useful information.

39.   All three residual plots (below) show evidence of curvature, indicating that higher-order terms (e.g., x_1^2, x_2^2) should be considered as additions to the model.

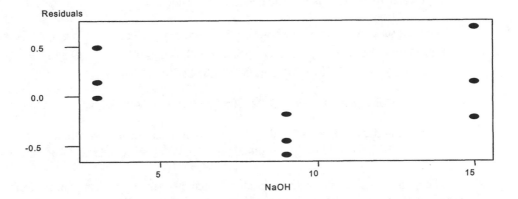

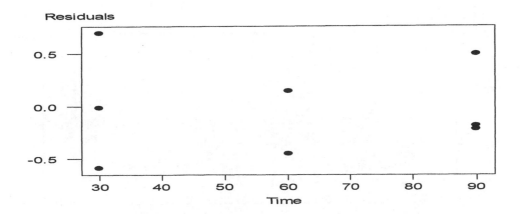

Residuals

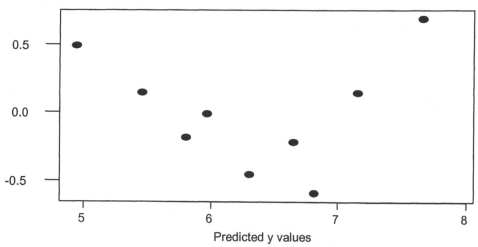

Predicted y values

41. (a) Exactly one car (x =1) and exactly one bus (y = 1) appear at the intersection .030 (i.e., 3%) of the time.

(b) At most one vehicle of each type (x ≤ 1 and y ≤ 1) appear at the intersection: p(0,0) + p(0,1) + p(1,0) + p(1,1) = .025 + .015 + .050 + .030 = .120 or, about 12% of the time.

(c) The number of cars equals the number of buses (x = y): p(0,0) + p(1,1) + p(2,2) = .025 + .030 + .050 = .105 or, about 10.5% of the time.

(d) Adding the rows yields the marginal distribution of the x values:

x :	0	1	2	3	4	5
p(x):	.05	.10	.25	.30	.20	.10

so, the mean number of cars is μ_x = 0(.05) + 1(.10) + 2(.25) + 3(.30) + 4(.20) + 5(.10) = 2.8.

(e) The number of vehicle spaces occupied is h(x,y) = x+3y. The mean number of spaces occupied is then, $\mu_{h(x,y)}$ = $\sum\sum(x+3y)p(x,y)$ = [0 + 3(0)](.025) + [0+3(1)](.015) + ...+ [5+3(2)](.020) = 4.90.

43. Summing across the rows (for x) and the columns (for y), the marginal distributions of x and y are:

x: 200 500   y: 0 250 500
   .50 .50   .25 .25 .50

So, the means of the two distributions are: μ_x = 200(.50) + 500(.50) = 350 and μ_y = 0(.25) + 250(.25) + 500(.50) = 312.5. The covariance between the two variables is then:

Covariance(x,y) = $\sum\sum(x-\mu_x)(y-\mu_y)f(x,y)$, where f(x,y) is the joint mass function given in exercise 38. That is, covariance(x,y) = (200-350)(0-312.5)(.20) + (200-350)(250-312.5)(.10)+...+ (500-350)(500-312.4)(.30) = 9375.0.

Using the marginal distributions again, the standard deviations of each variable can be computed:

$\sigma_x^2 = \sum(x-\mu_x)^2 p(x)$ = (200-350)^2(.50) + (500-350)^2(.50) = 22500, so σ_x = 150 and,

$\sigma_y^2 = \sum(y-\mu_y)^2 p(y)$ = (0-312.5)^2(.25) + (250-312.5)^2(.25)+ (500-312.5)^2(.50) =

42,968.75, so σ_y = 207.289. The population correlation coefficient is then ρ = covariance/($\sigma_x\sigma_y$) = 9375/[(150)(207.289)] = .302. Because it is closer to 0 than to 1, this value of ρ does not suggest that there is a strong relationship between x and y.

45.　(a) b = .500 means that, on average, for each 1 °F increase in temperature there is about .500 unit increase in strength.

(b) The least squares line is $\hat{y}$ = -25.000 + .500x, so for x = 120, $\hat{y}$ = -25.000 + .500(120) = 35. The residuals for the values y = 40 and y = 29 are: y- $\hat{y}$ = 40-35 = 5 and y- $\hat{y}$ = 29-35 =-6. The residuals have different signs because one point (x=120, y=40) lies above the regression line and the other (x=120, y=29) lies below the line.

(c) Coefficient of determination = r^2 = 1 - SSResid/Ssto = 1 - 390.0/1060.0 = .632, or about 63.2%. This means that about 63.2% of the observed variation in strength can be attributed to the approximate linear relationship between strength and treatment temperature. This is a good value for r^2 (it is closer to 1 than to 0), but it might be possible to improve/increase r^2 by including some additional predictor variables and fitting a multiple regression model.

47.　(a) Since stride rate is being predicted, y = stride rate and x = speed. Therefore, SS_{xx} = $\sum x_i^2 - (\sum x_i)^2/n$ = 3880.08 - (205.4)^2/11 = 44.7018, $SS_{yy} = \sum y_i^2 - (\sum y_i)^2/n$ = 112.681 - (35.16)^2/11 = .2969, and $SS_{xy} = \sum x_i y_i - (\sum x_i)(\sum y_i)/n$ = 660.130 - (205.4)(35.16)/11 = 3.5969. Therefore, b = SS_{xy}/SS_{xx} = 3.5969/44.7018 = .0805 and a = (35.16/11) - (.0805)(205.4/11) = 1.6932. The least squares line is then $\hat{y}$ = 1.6932 + .0805x.

(b) Predicting speed from stride rate means that y = speed and x = stride rate. Therefore, interchanging the x and y subscripts in the sums of squares computed in part (a), we now have SS_{xx} = .2969 and SS_{xy} = 3.5969 (note that SS_{xy} does not change when the roles of x and y are reversed). The new regression line has a slope of b = SS_{xy}/SS_{xx} = 3.5969/.2969 = 12.1149 and an intercept of a = (205.4/11) - (12.1149)(35.16/11) = -20.0514; that is, $\hat{y}$ = -20.0514 + 12.1149x.

(c) For the regression in part (a), $r = 3.5969/[\sqrt{44.7018}\sqrt{.2969}\,] = .9873$, so $r^2 = (.9873)^2$
= .975. For the regression in part (b), r is also equal to .9873 (since reversing x and y
has no effect on the formula for r). So, both regressions have the same coefficient of
determination. For the regression in part (a), we conclude that about 97.5% of the
observed variation in rate can be attributed to the approximate linear relationship
between speed and rate. In part (b), we conclude that about 97.5% of the variation in
speed can be attributed to the approximate linear relationship between rate and speed.

49. (a) Substituting x = .005 into the least squares equation, a stress value of $\hat{y} = 88.791$
     $+ 5697.0(.005) - 328,161(.005)^2 = 109.07$ would be predicted.

(b) From the stress values (i.e, y values given in the exercise,) $SSТо = SS_{yy} = \sum y_i{}^2$
    $- (\sum y_i)^2/n = 107,604 - (1034)^2/10 = 688.40$. Subtracting predicted values from
    actual values gives the residuals: -3.16, -1.87, 5.07, 1.93, 2.84, -3.36, -1.48, -2.22,
    2.07, and 0.20. The sum of squares of these residuals is $SSResid = 73.711$. The
    coefficient of determination $r^2 = 1 - SSResid/SSTo = 73.711/688.40 = .893$. Therefore,
    about 89.3% of the observed variation in stress can be attributed to the approximate
    linear relationship between stress sand strain.

(c) Substituting x = .03 into the least squares equation yields a predicted stress value of
    $\hat{y} = 88.791 + 5697.0(.03) - 328,161(.03)^2 = -35.64$, which is not at all realistic (since
    stress values can not be negative). The problem is that the value x = .03 lies far outside
    the region of x values that were used to fit the regression equation (the maximum x value
    used was x = .017). Extrapolation such as this is often unreliable (at best) and
    sometimes leads to ridiculous (i.e., impossible) predictions as in this case.

51. (a) The curvature that is apparent in the plot of y versus x (see below) indicates that merely
     fitting a straight-line to the data would not be the best strategy. Instead, one should search
     for some transformation of the x or y data (or both) that would give a more linear plot.

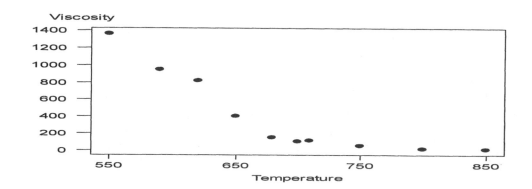

(b) The plot below shows the graph of ln(y) versus 1/x. Because it appears to be approximately linear, a straight-line fit to such data should provide a reasonable approximation to the relationship between the two variables.

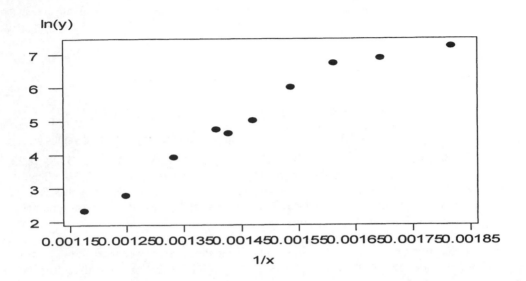

The following Minitab printout shows the results of fitting a regression line to the transformed data. From the printout, the prediction equation is $\ln(y) \approx -7.2557 + 8328.4(1/x)$. The r^2 value of 95.3% indicates that the fit is quite good. When temperature is 720 (i.e., $x = 720$), the equation gives a predicted value of $\ln(y) \approx -7.2557 + 8328.4(1/720) = 4.31152$. Exponentiating both sides gives a predicted y value of $y \approx e^{4.31152} = 74.6$.

```
The regression equation is
logey = - 7.26 + 8328 recipx

Predictor        Coef         StDev            T          P
Constant      -7.2557        0.9670        -7.50      0.000
recipx         8328.4         651.1        12.79      0.000

S = 0.3882      R-Sq = 95.3%      R-Sq(adj) = 94.8%

Analysis of Variance

Source         DF          SS            MS          F          P
Regression      1        24.663        24.663     163.63     0.000
Error           8         1.206         0.151
Total           9        25.869
```

56

53.   A plot of the data shows a nonlinear relationship that is similar in shape to Pattern 3 on page 130 of the text.

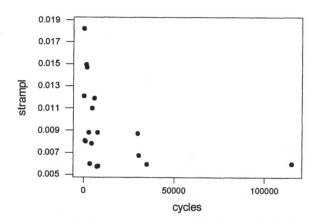

According to the guidelines on that page, the list of possible transformations for the x (*and* y) variables are: $\sqrt{x}, \sqrt[3]{x}, \ln(x), \frac{1}{x}$.   Below is a graph of ln(y) versus ln(x):

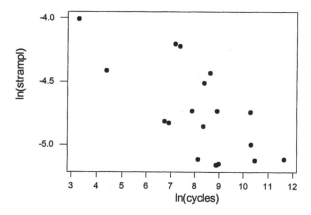

This graph appears to fairly linear and the variation in the ln(y) values appears to be reasonably constant over all values of x, so we can fit a simple linear regression model to this data. A Minitab printout of the resulting regression analysis is given below:

```
The regression equation is
lnstramp = - 3.74 - 0.124 lncycles

Predictor          Coef        StDev          T          P
Constant        -3.7372       0.2695      -13.87      0.000
lncycles        -0.12395      0.03199       -3.87      0.001

S = 0.2731       R-Sq = 46.9%      R-Sq(adj) = 43.8%

Analysis of Variance

Source           DF            SS           MS          F          P
Regression        1         1.1190       1.1190      15.01      0.001
Residual Error   17         1.2675       0.0746
Total            18         2.3865
```

By exponentiating both sides of the estimated regression model $\ln(y) = -3.7372 - 0.12395\ln(x)$, the model can be rewritten in terms of the original x and y variables:

$$y = e^{\ln(y)} = e^{-3.7372 - .12395\ln(x)} = (e^{-3.7372})(e^{-.12395\ln(x)}) = .02382\, x^{-.12395} \ .$$

For 5000 cycles to failure, the predicted amplitude is:

$$y = .02382\,(5000)^{-.12395} = .00829.$$

Chapter 4
Obtaining Data

1.   Operational definitions are used to define measurement procedures. Benchmarks are objects or procedures used to compare two or more products or processes.

3.   For example, you might use a definition such as: *Temperature at 2:00 pm in a fixed, unshaded area on top of City Hall.*

5.   An obvious method would be: *Obtain an exact count of the number N of words in the dictionary. Use a random number generator (e.g., in Excel or a statistical software package) to select a random sample of n integers from the numbers 1, 2, 3, ..., N.* Although this sounds simple, finding N is usually not easy. For example, should you count singular and plural forms of a word as being the same or different words? Some operational definition will have to be created so that you know exactly what words to count when finding N, and even after doing so, finding N will be a huge task so this is generally not a practical method.

    There are many methods that may sound good, but do not actually produce random samples. For instance, one such method is: *Pick a page at random, then pick a column at random (assuming there are 2 columns per page), then pick a line at random from the column, then use the word (in bold) that is described by that line.* This method does not produce truly random selections because it gives words with several meanings more chance of being selected than words with few meanings. Even if you alter the method to: *Pick a word at random from the randomly selected page*, words with several meanings will still have a greater probability of being selected.

7   (a) Yes; because any random sample from the complement of a random sample can be added to the original sample (Rule 4, page 160).

    (b) Take a random sample of 15 from the original random sample of 20 (Rule 2, page 60).

9.   Homogeneity within strata increases the precision (i.e., reduces the standard deviation) of estimates derived from the data. The goal in selecting strata is to try to have a lot of <u>similarity</u> between the elements *within* a stratum, but to have <u>dissimilarity</u> *between* strata.

11. (a) Weighted average of sample proportions = $\sum_i \left(\frac{N_i}{N}\right) p_i$

(b) Since each $n_i = (N_i/N)n$, then

$$\sum_i \left(\frac{N_i}{N}\right) p_i = \sum_i \left(\frac{N_i}{N}\right)\left(\frac{x_i}{n_i}\right) = \sum_i \left(\frac{N_i}{N}\right)\left(\frac{x_i N}{n N_i}\right) = \sum_i \frac{x_i}{n} = \sum_i x_i / \sum_i n_i$$

That is, when you use *proportional allocation*, you can simply treat the n data points $(x_1, x_2, \ldots, x_n)$ as if they are a single sample of size n, using $x_1 + \ldots x_n$ as the total number of objects having the particular characteristic out of a sample of size $n = n_1 + \ldots + n_k$. Because of this desirable property, samples based on proportional allocation are often called 'self-weighting'. The same self-weighting property holds when combining means from k samples when proportional allocation is used to obtain each of the k samples.

13. Note to work this problem, you need to use the formula presented in problem 14 of the text (for the case of $k = 2$ strata):

$$w_i = \frac{\frac{N_i \sigma_i}{\sqrt{c_i}}}{\frac{N_1 \sigma_1}{\sqrt{c_1}} + \frac{N_2 \sigma_2}{\sqrt{c_2}}} \quad \text{for } i = 1 \text{ and } 2$$

From the problem we know that $.20 = N_1/(N_1+N_2)$, $.80 = N_2/(N_1+N_2)$, $\sigma_1 = \sigma_2$, and $c_1 = 2c_2$. The fact that $\sigma_1 = \sigma_2$ allows us to cancel σ_1 and σ_2 from top and bottom of the formula for w_i, so:

$$w_i = \frac{\frac{N_i}{\sqrt{c_i}}}{\frac{N_1}{\sqrt{c_1}} + \frac{N_2}{\sqrt{c_2}}} \quad \text{for } i = 1 \text{ and } 2$$

Dividing top and bottom through by N and then replacing N_1/N by .2 and N_2/N by .8, the expression simplifies to:

$$w_1 = \frac{\frac{.2}{\sqrt{c_i}}}{\frac{.2}{\sqrt{c_1}} + \frac{.8}{\sqrt{c_2}}} \quad \text{and} \quad w_2 = \frac{\frac{.8}{\sqrt{c_2}}}{\frac{.2}{\sqrt{c_1}} + \frac{.8}{\sqrt{c_2}}}$$

Finally, substituting $2c_2$ for c_1 gives:

$$w_1 = \frac{\frac{.2}{\sqrt{2c_2}}}{\frac{.2}{\sqrt{2c_2}} + \frac{.8}{\sqrt{c_2}}} = \frac{\frac{.2}{\sqrt{2}}}{\frac{.2}{\sqrt{2}} + \frac{.8}{\sqrt{1}}} = .15022 \quad \text{and} \quad w_2 = \frac{\frac{.8}{\sqrt{c_2}}}{\frac{.2}{\sqrt{2c_2}} + \frac{.8}{\sqrt{c_2}}} = \frac{\frac{.8}{\sqrt{1}}}{\frac{.2}{\sqrt{2}} + \frac{.8}{\sqrt{1}}} = .84978.$$

Therefore, the allocation should be:

$n_1 = w_1 n = .15022(1000) = 150.22$ and $n_2 = w_2 n = .84978(1000) = 849.78.$

Rounding these results to integer values gives $n_1 = 150$ and $n_2 = 850$.

15.   In order to estimate each stratum's variance by the sample standard deviation s_i, the absolute minimum that <u>each</u> n_i can be is 2 (it takes at least two observations to estimate a variance). Therefore, the total sample size must satisfy the inequality:

$$n = n_1 + n_2 + \ldots + n_k \geq 2 + 2 + \ldots + 2 = 2k \quad (\text{i.e, } n \geq 2k), \quad \text{and so } k \ \underline{\text{cannot}} \text{ exceed } n/2.$$

17.   Use people as blocks; i.e., select a random sample of people and have each person test both packages.

19.   (a) More photoresist levels and fewer replications.

        (b) More replication and fewer photoresist levels.

21.   (a) Neither design is preferable; both involve some randomization of the assignment of mixes to subplots.

        (b) Experiment B is preferred: use large subplots as blocks and test each mix on each large subplot.

23.   (a) On average, there is an increase in hardness of $(3.2+3.6)/2 - (2.6+2.8)/2 = 0.7$ when switching from Brand A to B..

        (b) On average, there is an increase in hardness of $(2.8+3.6)/2 - (2.6+3.2)/2 = 0.3$ when switching from Machine 1 to 2.

        (c) Add some replicate observations in each cell of the experiment.

25.   Calibration has *no* effect on the standard deviation of a set of measurements.

27.   (a) No; the entire experiment must be replicated, not just the final measurements.

        (b) The standard deviation measures the variation in the readings given by the electronic balance.

29.   Most importantly, ask what '30%' means. Does it mean 30% of the area will receive rain, or that 30% of all such conditions in the past lead to rainy days, etc.

31.   (a) Setting the proportions equal gives $5/50 \approx 100/T$, or , $T \approx 1{,}000$.

        (b) $x_{tag2}/y \approx x_{tag1}/T$, which yields $T \approx x_{tag2}/(yx_{tag1})$.

33.   (a)

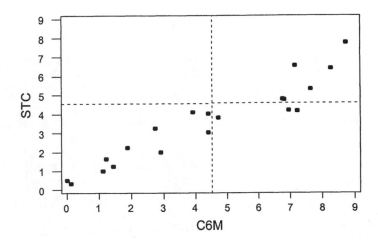

        (b) Regression line $STC = 0.525 + 0.686 \ C6M$.

        (c) The plot in (a) shows that there is some consistency within each lab, but that most
             labs are following different procedures than others. The regression line does not
             equal 1, so the different labs also appear to be giving systematically biased
             measurements.

Chapter 5
Probability and Sampling Distributions

1.   (a) Sampling without replacement means that no repeated items will occur in any sample. There are 10 possible such samples of size 3:

{a,b,c}, {a,b,d}, {a,b,e}, {a,c,d}, {a,c,e}, {a,d,e}, {b,c,d}, {b,c,e}, {b,d,e}, {c,d,e}.

As you learned in the discussion of the binomial distribution, the number of ways to choose a sample of 3 distinct items from a list of 5 items is given by $\binom{5}{3} = 10$, which shows that the list above is indeed complete.

   (b) Event A contains 3 samples from the list in (a); i.e., A = { {a,b,c}, {a,c,d}, {a,c,e}}

   (c) The complement of A is
        A′ = {{a,b,d}, {a,b,e}, { {a,d,e}, {b,c,d}, {b,c,e}, {b,d,e}, {c,d,e}}

3.   To envision the events A and B, it helps to draw a number line with the integers representing the possible numbers of defective items:

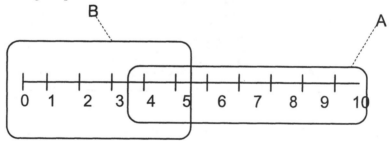

   (a) The event *A and B* consists of the integers {4, 5}. That is, *A and B* is the event 'either 4 or 5 defectives in the sample'.

   (b) The event *A or B* consists of all ten integers. There are many ways to describe this event. One description is that *A or B* is the event 'there is at least one defective in the sample'.

   (c) The complement of A consists of the integers {0, 1, 2, 3}. In words, A′ is the event 'there are at most 3 defectives in the sample'.

5.   The tree diagram is shown below. Note that it is not necessary for all the branches to have the same length. That is, some branches may stop early in the tree, while others may extend through several additional branching points.

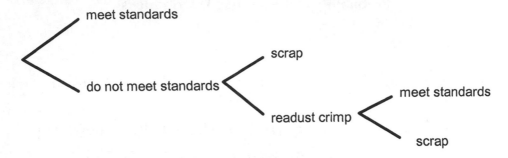

7.   The event *A and B* is the shaded area where A and B overlap in the following Venn diagram. Its complement consists of all events that are either not in A or not in B (or not in both). That is, the complement can be expressed as *A′ or B′*.

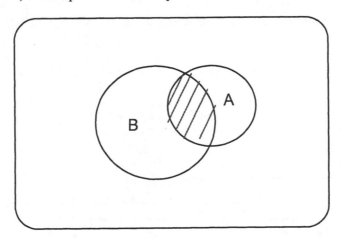

9.   (a) The total of 724 + 751 solder joints overstates the actual number found, since 316 of solder joints were found by *both* inspectors. To avoid such double-counting (i.e., the 316 is part of both the 724 joints found by inspector A and the 751 joints found by Inspector B), 316 should be subtracted from the raw totals, which means that 724 + 751 − 316 = 1,159 *distinct* joints were identified by the inspectors together. The important point to note in this problem is that the events 'Inspector A finds a defective solder joint' and 'Inspector B finds a defective solder joint' are *not* necessarily mutually exclusive, so we can not simply add the numbers of joints (or, equivalently, the probabilities of finding defective joints) for both inspectors.

        (b) Event *A and B′* contains 724 − 316 = 408 solder joints.

11.   Letting A_i denote the event that the ith component fails (note that this is different from the definition of A_i used in problem 5.10), the probability that the entire series system fails is denoted by $P(A_1$ or A_2 or A_3 or ... or $A_k)$. Given that each $P(A_i) = .01$, the problem states that $P(A_1$ or A_2 or A_3 or ... or $A_k) \leq P(A_1) + P(A_2) + P(A_3) + ... + P(A_k) = .01 + .01 + ... + .01 = 10(.01) = .10$. That is, there is *at most* a 10% chance of system failure.

13.   (a) The events *A and B* and *A and B'* are depicted by separate paths in Figure 5.5, so they mutually exclusive events. Furthermore, the diagram shows that A occurs when *either A and B* occurs or when *A and B'* occurs. Therefore, the addition law for mutually exclusive events allows us to write: $P(A) = P(A$ *and B* <u>or</u> *A and B'*$) = P(A$ *and B* $) + P(A$ *and B'*$)$.

       (b) Since $P(A|B) = P(A$ *and B*$)/P(B)$, multiplying through by $P(B)$ give the equivalent formula $P(A$ *and B*$) = P(A|B) \cdot P(B)$. Similarly, $P(A|B') = P(A$ *and B'*$)/P(B')$ can be rewritten as $P(A$ *and B'*$) = P(A|B') \cdot P(B')$.

       (c) $P(B|A) = P(B$ *and A*$)/P(A)$ by the conditional probability formula. Note that *B and A* is the same event as *A and B*, so $P(B$ *and A*$) = P(A$ *and B*$)$. Next, substitute the expressions for $P(A$ *and B*$)$ and $P(A$ *and B'*$)$ from (b) into the formula for $P(A)$ in part (a). Similarly, substitute the expression for $P(A$ *and B*$)$ from (b) in for $P(B$ *and A*$)$. All of these substitutions result in the formula known as Bayes' formula or Bayes' theorem.

       (d) For example, going along the top branch results in the event *A and B*. The two branches comprising this event/path have probabilities $P(B)$ and $P(A|B)$. From (b) we know that $P(A$ *and B*$) = P(A|B) \cdot P(B)$. In other words, the ability to simply multiply probabilities of branches in a tree diagram is just a restatement of the basic conditional probability formula.

       (e) A simple way to find Bayes' formula would be to write the basic conditional probability formula $P(B|A) = P(B$ *and A*$)/P(A)$ and then use the observation in (d) to find other formulas for $P(B$ *and A*$)$ and $P(A)$. For example, the tree diagram shows that A can be decomposed into the two branches *A and B* and *A and B'*, and the multiplication of branches property from (d) allows you to immediately write down the denominator of Bayes' formula. Similarly, $P(B$ *and A*$)$ can immediately be written as then product of the corresponding branches in the tree diagram, giving the numerator of Bayes' formula.

15.   The probabilities are shown on the following tree diagram:

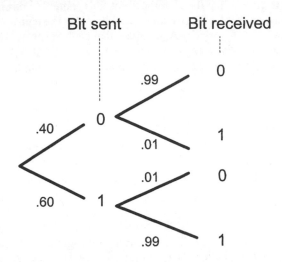

(a) The proportion/percentage of 1s received is sum of the proportion of 0s that were sent and mistakenly received as 1s, $(.40)(.01)$, plus the proportion of 1sd that were sent and correctly received as 1s, $(.60)(.99)$; i.e., $(.40)(.01) + (.60)(.99) = .04 + .594 = .634$, or 63.4%.

(b) Let S denote the event that a 1 is sent and let R be the event that a 1 is received. Then the problem asks for the conditional probability $P(S \mid R)$. The conditional probability formula gives $P(S \mid R) = P(S \text{ and } R)/P(R)$. Part (a) gives $P(R) = .634$. From the tree diagram $P(S \text{ and } R)$ is the product of the probabilities $(.60)(.99) = .594$. Therefore, $P(S \mid R) = .594/.634 = .937$, or 93.7%.

17.   (a) P(does not match the blood type) $= 1 - P(\text{have this blood type}) = 1 - \gamma$

(b) Random sampling assures that any person's chance of matching the blood type is <u>independent</u> of any other person's in the sample, so the probability that n people do not match the blood type is just $(1-\gamma)^n = P(\text{first person doesn't match}) \cdot P(\text{second person doesn't match}) \cdots P(n^{th} \text{ person doesn't match})$.

(c) Using the result from part (b), P(at least one match) $= 1 - P(\text{no matches}) = 1 - (1-\gamma)^n$.

(d) Using the result from part (c), for $\gamma = 10^{-6}$, $1 - (1-\gamma)^n = 1 - (1-10^{-6})^{10^6} = .6321$.

19.   (a) P(both bids are successful) $= P(E_1 \text{ and } E_2) = P(E_1) \cdot P(E_2) = (.4)(.3) = .12$.

(b) P(neither bid is successful) $= P(E_1' \text{ and } E_2') = P(E_1') \cdot P(E_2') = (1-.4)(1-.3) = .42$. Recall that if two events are independent, so are their complements.

(c) P(at least one bid is successful) $= 1 - P(\text{neither is successful}) = 1 - P(E_1' \text{ and } E_2') = 1 - .42 = .58$.

21.   (a) Let R_i denote the event that the ith relay correctly transmits a bit. Therefore, $P(R_i)$
= .80 for each relay. If a one is sent initially, then the probability that all three relays
correctly send a 1 is $P(R_1$ and R_2 and $R_3)$. Because the relays are independent,
$P(R_1$ and R_2 and $R_3)$ = $P(R_1) \cdot P(R_2) \cdot P(R_3) = (.80)^3 = .512$.

(b) Note that it is possible for a 1 to be sent, then incorrectly transmitted as a 0 and then
(again incorrectly) converted back to a 1. That is, if two relays both malfunction, a 1
could still end up being *correctly* transmitted as a 1. The best way to keep track of such
eventualities is to draw a tree diagram. Note that the probabilities of .80 and .20 change
around on the various branches since they are probabilities of correct and incorrect
transmissions, not probabilities of 1s and 0s (to simplify the diagram, only the branches
with probabilities of .80 are noted; the remaining branches have probabilities of .20).

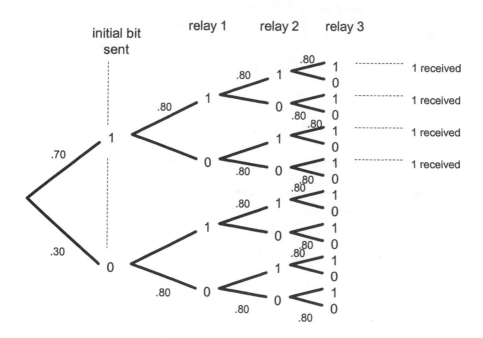

When 1 is sent initially, the upper half of the diagram applies. Note that the end branches
that result in a 1 being received are indicated on the diagram. Since there are 4 branches (in
the upper half of the diagram) that end with a 1, there are 4 probabilities to add (recall that
branches are mutually exclusive, so their probabilities add). Using the fact that branch
probabilities are simply the products of their sub-branch probabilities, we find:

P(1 is received when a 1 is sent) = $(.80)^3 + (.80)(.20)^2 + (.20)^2(.80) + (.20)(.80)(.20) = .608$.

(c) Let R denote the event that a 1 is received and let S be the event that a 1 is sent. Then, $P(S\,|\,R) = P(S \text{ and } R)/P(R)$. Using the conditional probability formula again, $P(S \text{ and } R)$ can be written as $P(R\,|\,S)\cdot P(S)$. In part (b) we found that $P(R\,|\,S) = .608$. Also, $P(S) = .70$, so $P(S \text{ and } R) = (.608)(.70) = .4256$. To find $P(R)$, break the event R into two exclusive events $R \text{ and } S$ and $R \text{ and } S'$. Note that $P(R\,|\,S')$ can be found in the same manner that $P(R\,|\,S)$ was found in (b): by following through the lower half of the tree diagram, $P(R\,|\,S') = (.20)(.80)^2 + (.20)^3 + (.80)(.20)(.80) + (.80)^2(.20) = .392$. Therefore, $P(R \text{ and } S') = P(R\,|\,S')\cdot P(S') = (.392)(.30) = .1176$. Finally, $P(R) = P(R \text{ and } S) + P(R \text{ and } S') = .4256 + .1176 = .5432$ and we find $P(S\,|\,R) = P(S \text{ and } R)/P(R) = .4256/.5432 = .7835$.

23. (a) Let S_i denote the event that the ith point signals a problem with the manufacturing process. Then, the probability that *none* of the 10 points give such a signal is P(no signals) $= P(S_1' \text{ and } S_2' \text{ and } ... S_{10}') = P(S_1')\cdot P(S_2')\cdot ...P(S_{10}') = (1-.01)^{10} = .90438$. Therefore, the probability of having *at least one* point signal a problem is P(at least one signal) = 1 - P(no signals) $= 1 - .90438 = .0956$.

(b) P(at least one in 25 signals a problem) $= 1 - (1-.01)^{25} = .2222$.

25. (a) Discrete; because x has a finite, countable number of values.

(b) Continuous; because concentrations can have any conceivable value in an interval.

(c) Discrete; the total number of bolts is finite, which means that the possible number of oversized bolts as well as the proportion of oversized bolts can only have a finite number of possible values.

(d) Discrete; the number of errors must be finite.

(e) Continuous; because strength measurements can conceivably have any value in an interval of real numbers.

(f) Continuous; because time can take any value in an interval of real numbers.

(g) Discrete; because the number of customers must be finite.

27. (a) x is a discrete random variable, so $\mu = \sum_x xp(x) = (1)(.2) + (2)(.4) + (3)(.3) + (4)(.1) = 2.3$.

(b) $\sigma^2 = \sum_x (x - \mu)^2 p(x) = (1-2.3)^2(.2) + (2-2.3)^2(.4) + (3-2.3)^2(.3) + (4-2.3)^2(.1) = .81$.

(c) The expected number of pounds shipped should be 5μ; i.e., the product of (5 lbs./order) times the expected number of orders shipped, μ. Therefore, the expected number of pounds remaining after an order is shipped is $100 - 5\mu = 100 - 5(2.3) = 88.5$ lbs.

29. (a) y is a discrete random variable, so $\sum_y p(y)$ must equal 1. Using the formula for

p(y), we have $k(1) + k(2) + k(3) + k(4) + k(5) = 1$, or, $15k = 1$, so $k = 1/15$.

(b) P(at most three forms required) $= P(y \le 3) = p(1) + p(2) + p(3) = 1(1/15) + 2(1/15) + 3(1/15) = 6/15 = 2/5 = .40$, or 40%.

(c) $\mu = \sum_y yp(y) = 1(1/15) + 2(2/15) + 3(3/15) + 4(4/15) + 5(5/15) = 55/15 = 11/3$.

(d) $\sigma^2 = \sum_y (y-\mu)^2 p(y) = (1-11/3)^2(1/15) + (2-11/3)^2(2/15) + (3-11/3)^2(3/15) +$

$(4-11/3)^2(4/15) + (5-11/3)^2(5/15) = 14/9 = 1.5555$. Therefore, $\sigma = \sqrt{14/9} = 1.2472$.

31. (a) x = 'number of defective solder joints' has a binomial mass function with $n = 285$ and $\pi = .01$, so $\mu = np = 285(.01) = 28.5$ and $\sigma = \sqrt{n\pi(1-\pi)} = 1.6797$.

(b) $P(x = 0) = \binom{285}{0}(.01)^0(1-.01)^{285} = (1-.01)^{285} = .05702$

(c) $P(x \ge 2) = 1 - P(x < 2) = 1 - [P(x=0) + P(x = 1)]$

$= 1 - [\binom{285}{0}(.01)^0(1-.01)^{285} + \binom{285}{1}(.01)^1(1-.01)^{284}] = 1 - [.05702 + .164150] =$

.77883.

33. (a) Let x = 'number of breakthroughs'. Then x has a binomial mass function with $n = 100$ and $\pi = .50$, so $\mu = np = 100(.50) = 50$.

(b) Using the normal approximation (with continuity correction) to binomial along with the fact that $\sigma = \sqrt{n\pi(1-\pi)} = 5$ gives $P(x \ge 60) \approx P(z \ge \dfrac{59.5-50}{5}) = P(z \ge 1.90) = 1 - P(z \le 1.90) = 1 - .9713 = .0287$.

35. (a) x = 'number of defectives in a sample' has a binomial mass function with n = 10 and π = .10. Therefore, using Appendix Table II (page 562), P(accepting a lot) = P(x ≤ c) = P(x ≤ 1) = .349 + .387 = .736, so the probability of rejecting the lot is 1- .736 = .264.

(b) For a shipment with no defectives, π = 0, so the number of defectives in any sample must be x = 0, so it is a certainty that x ≤ 1; that is, P(accepting such a lot) = P(x ≤ 1) = 1.

(c)

π	P(x ≤ 1) = P(x=0) + P(x =1)	P(reject lot) = 1 - P(x ≤ 1)
.05	.914 = .599 + .315	.086 = 1 - .914
.20	.375 = .107 + .268	.625 = 1 - .375
.50	.011 = .001 + .010	.989 = 1 - .011

(d) The following Minitab plot shows the OC curve for a sampling plan with n =10, c = 1:

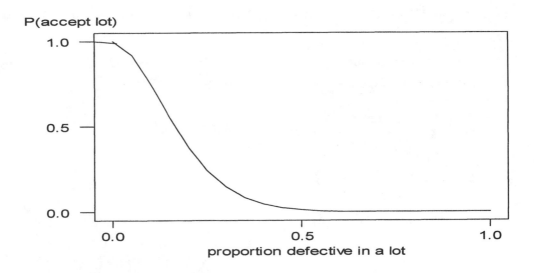

70

37. (a) x = 'number of correct answers' has a binomial mass function with n = 25 and
         π = 1/5 = .20.

     (b) μ = 25(1/5) = 5 and σ = $\sqrt{n\pi(1-\pi)}$ = 2.

     (c) Using Table II (page 564), the closest integer score S that satisfies P(x ≥ S) = .01 is S
         = 11 for which P(x ≥ 11) = .004 +.002 = .006.

39. (a) x = 'lifetime of an electronic component' has an exponential density with density function
         f(x) = $\lambda e^{-\lambda x}$ (for x ≥ 0). In this problem λ = 1/μ =1/MTBF = 1/500. Therefore, the
         median $\tilde{\mu}$ of this random variable satisfies the equation:

         $$.5 = \int_{0}^{\tilde{\mu}} \lambda e^{-\lambda x} dx = \left[-e^{-\lambda x}\right]_{0}^{\tilde{\mu}} = -[1 - e^{-\lambda\tilde{\mu}}].$$ That is, -.5 = 1 - $e^{-\lambda\tilde{\mu}}$, so $e^{-\lambda\tilde{\mu}}$ =

         1-.5 = .5.   Taking natural logarithms of both sides gives -λ $\tilde{\mu}$ = ln(.5), or, $\tilde{\mu}$ = -ln(.5)/λ
         = .69315(500) = 346.57 hours.

     (b) From the answer to part (a), $\tilde{\mu}$ = .69315/λ = .69315μ, which is less than μ.

     (c) $\tilde{\mu}$ = .69315μ, or approximately, $\tilde{\mu}$ = .693 × MTBF.

41. (a) P(x = 5) = .20 + .15 + .05 = .40
         P(x = 6) = .10 + .15+ .10 = .35
         P(x = 7) = .10 + .10 + .05 = .25

     (b) P(y = 10) = .20 + .15 + .05 = .40
         P(y = 15) = .15 + .15 + .10 = .40
         P(y = 20) = .05 + .10 + .05 = .20

     (d) The (x,y) pairs that satisfy x + y ≤ 21 are: (5,10), (5,15), (6,10), (6,15), and (7,10), so
         P(x + y ≤ 21) = .20 + .15 + .10 + .15 + .10 = .95.

     (e) x and y are <u>not</u> independent because P(x = 5 and y = 10) = .20 , whereas P(x = 5)P(y =
         10)
         = (.40)(.40) = .16, so P(x=5 and y=10) ≠ P(x=5)P(x=10).

43. (a) x has a binomial distribution with n = 5 and π = .05. Writing P(.05-.01 ≤ p ≤ .05+.01)
         in terms of x, we find P(.05-.01 ≤ x/5 ≤ .05+.01) = P((.04)5 ≤ x ≤ (.06)5) =
         P(.2 ≤ x ≤ .3). Because x can only have integer values, there is no x between .2 and .3,
         so the probability of this event is 0.

71

(b) For n= 25, $P(.04 \le p \le .06) = P((.04)25 \le x \le (.06)25) = P(1 \le x \le 1.5) = P(x = 1) =$
$\binom{25}{1}(.05)^1(.95)^{24} = 0.36498$.

(c) For n= 1005, $P(.04 \le p \le .06) = P((.04)100 \le x \le (.06)100) = P(4 \le x \le 6)$. Using the
normal approximation to the binomial, with $\mu = n\pi = 100(.05) = 5$ and $\sigma^2 = n\pi(1-\pi) =$
$100(.05)(.95) = 4.75$ and $\sigma = 2.7195$: $P(4 \le x \le 6) \approx P(\dfrac{4-5}{2.7195} \le z \le \dfrac{6-5}{2.1795}) =$
$P(-.46 \le z \le .46) = .3544$. Using the *continuity correction* makes a substantial difference
in this problem because the interval from 4 to 6 contains the mean of the distribution
(and hence, a large amount of the probability): $P(4 \le x \le 6) \approx P(\dfrac{3.5-5}{2.7195} \le z \le \dfrac{6.5-5}{2.1795})$
$= P(-.69 \le z \le .69) = .5098$.

45. (a) x = 'disconnect force' has a uniform distribution on the interval [2,4]. M is the maximum
of a sample of size n = 2 from the uniform density on [2,4]. The larger of two items
randomly selected from the interval [2,4] should, on average, tend to be closer to the
upper end of the interval.

(b) Using the same reasoning as in part (a), the largest value in a sample of n = 100 will,
most likely, be even closer to the upper end of the interval [2,4] than is the largest value
in a sample of size n =2. So the average of all M's based on n=100 ought to be larger
than the average vale of all M's based on n = 2.

(c) For larger samples (e.g., n = 100), the maximum values will usually be fairly close to
upper endpoint of 4, which means that the variability amongst such values will tend to
be small. For smaller samples (e.g., n = 2), it is easier for the value of M to wander over
the interval [2,4], which means that the variability among these values will be larger
than for n = 100.

47. (a) $\mu_p = \pi = .80$. $\sigma_p = \sqrt{\dfrac{\pi(1-\pi)}{n}} = \sqrt{\dfrac{.80(1-.80)}{25}} = .08$

(b) Since 20% do <u>not</u> favor the proposed changes (so $\pi = .20$), the mean & standard
deviation of the sampling distribution of this proportion are $\mu_p = \pi = .20$ and
$\sigma_p = \sqrt{\dfrac{\pi(1-\pi)}{n}} = \sqrt{\dfrac{.20(1-.20)}{25}} = .08$.

(c) For n = 1000 and $\pi = .80$, $\mu_p = \pi = .80$. $\sigma_p = \sqrt{\dfrac{\pi(1-\pi)}{n}} = \sqrt{\dfrac{.80(1-.80)}{100}} = .04$. Notice that it was necessary to <u>quadruple</u> the sample size (from n=25 to n=100) in order to cut σ_p in half (from $\sigma_p = .08$ to $\sigma_p = .04$).

49.   (a) n = 16, $\mu = 12$ and $\sigma = .04$, so $P(11.99 \le \bar{x} \le 12.01) = P(\dfrac{11.99-12}{.04/\sqrt{16}} \le z \le \dfrac{12.01-12}{.04/\sqrt{16}}) =$

$P(-1 \le z \le 1) = .8413 - .1587 = .6826$.

(b) n = 64, $\mu = 12$ and $\sigma = .04$, so $P(11.99 \le \bar{x} \le 12.01) = P(\dfrac{11.99-12}{.04/\sqrt{64}} \le z \le \dfrac{12.01-12}{.04/\sqrt{64}}) =$

$P(-2 \le z \le 2) = .9772 - .0228 = .9544$.

51.   (a) x = 'lifetime of battery" has a normal density with $\mu = 8$ hours and $\sigma = 1$ hour.

Therefore, P(average of 4 exceeds 9 hours) = $P(\bar{x} > 9) = P(z > \dfrac{9-8}{1/\sqrt{4}}) = P(z > 2) =$

$1 - P(z<2) = 1 - .9772 = .0228$.

(b) Having the total lifetime of 4 batteries exceeds 36 hours is the same thing as having their average exceed 9, so the probability of this event is .0228, the same as in part (a).

(c) $.95 = P(T > T_0) = P(T/4 > T_0/4) = P(\bar{x} > T_0/4) = P(z > \dfrac{T_0/4-8}{1/\sqrt{4}}) = P(z > T_0/2-16)$.

For a standard normal distribution, $P(z > 1.645) \approx .95$, so we must have $T_0/2-16 = 1.645$, which gives $T_0 = 8 + 1.645/2 = 8.8225$ hours.

(d) Let Y = "replacement cost". Then Y is a discrete random variable with values $0 and $3 and corresponding probabilities .95 and .05. The expected value of Y (i.e., the average replacement cost per package) is $\mu = \$0(.95) + \$3(.05) = \$0.15$ per package.

53. (a) x = 'sediment density' has a normal distribution with $\mu = 2.65$ and $\sigma = .85$), so

$P(\bar{x} \le 3.00) = P(z \le \dfrac{3.00 - 2.65}{.85/\sqrt{25}}) = P(z \le 2.06) = .9803$. Similarly, $P(2.65 \le \bar{x} \le 3.00)$

$= P(0 \le z \le 2.06) = .9803 - .5000 = .4803$.

(b) For any n, $P(\bar{x} \le 3) = P(z \le \dfrac{3.00 - 2.65}{.85/\sqrt{n}}) = P(z \le .41176\sqrt{n})$. Since the value of

$z \approx 2.326$ is associated with a cumulative probability of .99, we equate $.41176\sqrt{n}$ and 2.326 and find $n = (2.326/.41176)^2 = 31.91$, or, about $n = 32$.

55. (a) Let p = 'proportion of resistors exceeding 105 Ω'. Then the sampling distribution of p is

approximately normal with $\mu_p = \pi = 0.02$ and $\sigma_p = \sqrt{\dfrac{\pi(1-\pi)}{n}} = \sqrt{\dfrac{.02(1-.02)}{100}} =$

0.014.

(b) $P(p < .03) = P(z < \dfrac{.03 - .02}{.014}) = P(z < .71) = 0.7611$.

57. (a) Let x = 'particle radius'. Then $y = \ln(x)$ has a normal distribution with $\mu = -2.62$ and
$\sigma = .788$, so $\mu_x = \exp(\mu + \sigma^2/2) = \exp(-2.62 + (.788)^2/2) = \exp(-2.3095) = 0.099308$.

(b) $P(x > .12) = P(\ln(x) > \ln(.12)) = P(y > -2.1203) = P(z > \dfrac{-2.1203 - (-2.62)}{.788}) = P(z > .63)$

$= 1 - P(z \le .63) = 1 - .7357 = 0.2643$.

59. (a) The sum of the parcel areas is $15+20+...+20 = 90$, so $P(B_1$ or B_2 or $B_3) = P(B_1) + P(B_2)$
$+ P(B_3) = 15/90 + 20/90 + 25/90 = 60/90 = 2/3$.

(b) $P(B_5') = 1 - P(B_5) = 1 - 20/90 = 7/9$.

61. Let B_i denote the event that the ith battery operates correctly. Therefore, $P(B_i) = .95$ for each
$i = 1,2,3,4$. Then, P(tool fails) = P(at least one battery fails) = 1 - P(all batteries operate
correctly) = $1 - P(B_1$ and B_2 and B_3 and $B_4) = 1 - P(B_1)P(B_2)P(B_3)P(B_4) = 1 - (.95)^4 = .1855$.

63. P(at least one event occurs) = P(A or B) = 1 - P(neither event occurs) = 1 - P(A' and B') =
$1 - P(A')P(B') = (1-P(A))(1-P(B))$. We have used the fact that A' and B' are independent to
simplify P(A' and B').

65.   (a) The total area under the density curve must equal 1, so:

$$1 = \int_{-\infty}^{\infty} f(x)\,dx = \int_0^b \tfrac{1}{2} x\,dx = \tfrac{1}{2}\left[\frac{x^2}{2}\right]_0^b = \frac{b^2}{4} \text{ and, therefore, } b^2 = 4 \text{ and } b = 2.$$

(b) $$\mu = \int_{-\infty}^{\infty} x f(x)\,dx = \int_0^b \tfrac{1}{2} x^2\,dx = \tfrac{1}{2}\left[\frac{x^3}{3}\right]_0^b = \frac{b^3}{6} = \frac{2^3}{6} = \frac{4}{3}.$$

(c) $$\sigma^2 = \int_{-\infty}^{\infty} (x-\mu)^2 f(x)\,dx = \int_0^b (x-\tfrac{4}{3})^2 \tfrac{1}{2} x\,dx = \tfrac{1}{2}\int_0^b (x^2 - \tfrac{8}{3}x + \tfrac{16}{9})x\,dx =$$

$$\tfrac{1}{2}\int_0^b (x^3 - \tfrac{8}{3}x^2 + \tfrac{16}{9}x)\,dx = \tfrac{1}{2}\left[\frac{x^4}{4} - \tfrac{8}{3}\left(\frac{x^3}{3}\right) + \tfrac{16}{9}\left(\frac{x^2}{2}\right)\right]_0^b = \frac{32}{243}, \text{ so } \sigma = .36289.$$

67.   (a) x is normal with $\mu = 3$ and $\sigma = .8$. Then, $P(x = 0) = 0$ because single 'points' such as $x = 0$ have zero probability when x is a continuous random variable.

(b) $P(x \geq 2) = P(z \geq \dfrac{2-3}{.8}) = P(z \geq -1.25) = 1 - P(z \leq -1.25) = 1 - .1056 = .8944.$

(c) $P(x > 2) = P(z \geq \dfrac{5-3}{.8}) = P(z \geq 2.5) = 1 - P(z \leq 2.5) = 1 - .9938 = .0062.$

69.   (a) x = 'battery voltage' has a mean value of $\mu = 1.5$ and a standard deviation of $\sigma = .2$ volts. The sampling distribution of $\bar{x}$ (based on n = 4) has a mean value of $\mu_{\bar{x}} = \mu = 1.5$ and a standard error of $\sigma_{\bar{x}} = \sigma/\sqrt{n} = .2/\sqrt{4} = .1$

(b) T is related to $\bar{x}$ by the equation $T = 4\bar{x}$, so the mean of T should be 4 times the mean of $\bar{x}$. That is, $\mu_T = 4\mu_{\bar{x}} = 4(1.5) = 6$. Similarly, the standard deviation of T should be 4 times that of $\bar{x}$, so $\sigma_T = 4\sigma_{\bar{x}} = 4(.1) = .4$

71.   (a)   Assuming that the births are <u>independent</u> of one another (i.e., the sisters didn't *plan* to have children as close to the same day as possible), then the probability of 3 births on the same day would be:

$$P(S_1 \text{ and } S_2 \text{ and } S_3) = P(S_1) \cdot P(S_2) \cdot P(S_3) = \left(\frac{1}{365}\right)^3.$$

(b)   For the births to simply occur on the same day, whether it is March 11 or another day, the probability is:

$P((\text{all 3 on Jan 01}) \text{ or } (\text{all 3 on Jan 02}) \text{ or } ....\text{or } (\text{all 3 on Dec 31})) =$

$P(\text{all 3 on Jan 01}) + P(\text{all 3 on Jan 02}) +...+ P(\text{all 3 on Dec 31})) =$

$$\left(\frac{1}{365}\right)^3 + \left(\frac{1}{365}\right)^3 +...+ \left(\frac{1}{365}\right)^3 = 365\left(\frac{1}{365}\right)^3 = \left(\frac{1}{365}\right)^2.$$

(c)   x = "Deviation of March 11 from the delivery date" is normal with $\sigma = 19.88$. For the sister with the March 15 due date, who is 4 days early in delivery, the corresponding probability under the normal curve is $P(3.5 < x < 4.5) \approx .0196$. Therefore, the probabilities are:

delivery date	deviation from mean	normal probability	
March 15	4 days early	$P(3.5 < x < 4.5)$	$\approx .0196$
April 1	21 days	$P(20.5 < x < 21.5)$	$\approx .0114$
April 4	24 days	$P(23.5 < x < 24.5)$	$\approx .0097$

Using independence, the probability that all three births occur on March 11 is the product: $(.0196)(.0114)(.0097) = 2.167 \times 10^{-6}$.

(d)   For a *common* birth date (not just March 11), you would repeat the calculations in part (c) for <u>each</u> of the 365 days of the year and then add all 365 results.

73.   (a)   The proportions of nonconforming cans, by production line, are:

$P(\text{line 1}) = 500/1500 = .33333$
$P(\text{line 2}) = 400/1500 = .26667$
$P(\text{line 1}) = 600/1500 = .40000$

So, there is a 33.33% chance that the selected can came from production line 1.

The total number of cracked cans is:
$.50(500) + .44(400) + (.40)(600) = 666$, so the probability that a nonconforming can is cracked is $666/1500 = .444$.

(b)   Restricting attention to Line 1, the table shows that 15% of all nonconforming cans have blemishes. So, the desired probability is simply 15%.

Alternatively, you could calculate that there are .15(500) = 75 nonconforming cans that are blemished from Line 1, so that P(Blemish and from Line 1) = 75/1500 = .05. <u>Given</u> that the defective can came from Line 1, then, the probability that it had a blemish is:

$$P(\text{Blemish} \mid \text{Line1}) = P(\text{Blemish and from Line 1}) / P(\text{Line 1}) =$$
$$.05/.33333 = .15.$$

(c)   P(Line 1 and Surface Defect) = .10(500)/1500 = .033333. Furthermore, the overall number of nonconforming cans with surface defects is: .10(500) +.08(400) + .15(600) = 172, which means P(Surface Defect) = 172/1500 = .1146667.

The desired probability is then:

$$P(\text{Line 1} \mid \text{Surface Defect}) =$$
$$P(\text{Line 1 and Surface Defect}) / P(\text{Surface Defect}) =$$
$$033333 / .1146667 = .2907.$$

Alternatively, you could simply calculate that there are 172 cans with surface defects and, since .10(500) = 50 of them come from Line 1, then the probability is 50/172 = .2907

Chapter 6
Quality Control

1.   The specification limits are 5% above and below the nominal value of 560. This tolerance is $\pm(.05)(560) = \pm 28$ Ohms, so LSL = 560-28 = 532Ω and USL = 560 + 28 = 588Ω.

2.   (a) The envelope puts an upper specification limit of 4.00 inches on the width of a folded letter.

      (b) Possible penalties: you may have to refold letter (rework), or bend letter to fit envelope (lower quality), or reprint and fold new letter (scrap and rework).

5.   (a) attributes data   (f) variables data
      (b) variables data   (g) attributes data
      (c) variables data   (h) variables data
      (d) attributes data   (i) variables data
      (e) attributes data

      One often-used rule for deciding whether a variable is continuous is the following: *if it is possible to obtain more and more precise measured values by using better and better measuring instruments, then the characteristic/variable is continuous (i.e, variables data).* Note that you don't have to actually obtain and use better instruments, you just have to perform the thought experiment of asking if instruments exist that will conceivably give more and more precise measurements.

7.   Some unacceptable parts whose *true* lengths are .02 inches or less below the LSL will give measured lengths above the LSL (and will then be incorrectly classified as acceptable). Conversely, some acceptable parts whose true lengths are less than .02 inches below the USL will have measured lengths above the USL (which incorrectly classifies them as unacceptable).

9.   Points within control limits should show random variation around the centerline with some points also falling near control limits. If all the points are packed closely around the centerline, then something is wrong; the data does not exhibit the random variation that is expected of sample statistics that follow (approximately) a normal distribution. Typical causes of this problem: the control limits may have been calculated incorrectly, measurements may have been rounded off too much, etc.

11. There will be <u>no effect on the R chart values</u>. To see this, let x_1, x_2, x_3,..., and x_n denote the *true* measurements in a subgroup of size n. Then, the altered measurements given by the instrument will be $x_1+\delta$, $x_2+\delta$, $x_3+\delta$,..., and $x_n+\delta$, so the range of the true measurements and the range of the altered measurements will be identical; i.e; $R_{true} = x_n-x_1 = (x_n+\delta)-(x_1+\delta) = R_{altered}$. Since the subgroup ranges for the altered data are the same as for the true values, there will be no changes in the average and centerline of the R chart.

However, the subgroup means of will differ by δ; i.e; the *true* subgroup mean is $\bar{x} = (x_1+...+x_n)/n$, while the mean of the altered data is $[(x_1+\delta)+(x_2+\delta)+(x_3+\delta)+...(x_n+\delta)]/n = \bar{x}+\delta$. The net effect will be that the centerline and the control limits of the x-bar chart will be shifted upwards by $+\delta$ units.

13. The A_2 factor decreases as n increases to the reflect the fact that the standard error of the sample mean, $\sigma/\sqrt{n}$ decreases as the subgroup size n increases. That is, as n increases, we expect the sampling distribution of the sample means to become more and more narrow (i.e., its standard error will become smaller and smaller, reflecting the fact that means based on large samples show less variation than do means based on smaller samples). The A_2 merely accounts for the different standard errors for the different sample sizes, so we do <u>not</u> expect a greater (or smaller) frequency of points outside the control limits for charts based on different subgroup sizes.

15. (a) A Minitab R chart for this data is shown below. The centerline is 0.0835, UCL = 0.2149, and LCL = 0.0000. There are no 'out of control' signals given by the chart. *Note: some of the titles on the chart were changed (using the Minitab graph editor) to more closely match the terminology in the text. Note also that Minitab use the expression '3.0SL' to stand for '3-sigma limit'; so the upper 3.0 SL is just the UCL and the lower 3.0 SL is the LCL.*

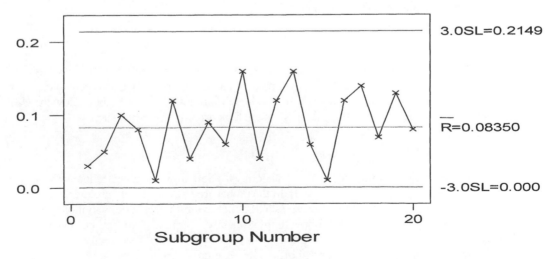

(b) A Minitab $\bar{x}$ chart is shown below. The centerline is 0.3813, UCL = 0.4668, and LCL = 0.2959. There are no 'out of control' signals given by the chart.

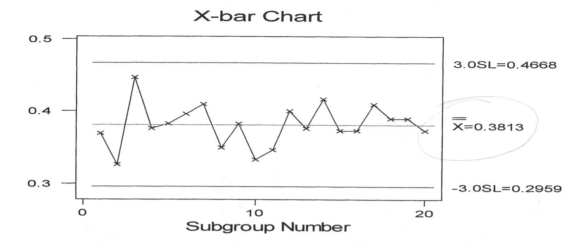

17. (a) A Minitab s chart of this data appears below. The centerline is 1.2642, UCL = $B_4\bar{s}$ = 1.970(1.2642) = 2.4905, and LCL = $B_3\bar{s}$ = (0.030)(1.2642) = 0.0379. *Note: to use Minitab to create this chart, you must first create 24 subgroups of size 6 whose subgroup means and standard deviations match those given in the exercise. An easy way to do this is to take any 6 numbers, stored them in a Minitab column, say C50, and then type the Minitab expression: MTB> let c1 =c51(1)+c52(1)*((c50-mean(c50))/stdev(c50)). Here, c51and c52 are columns containing the means and standard deviations given in the exercise. The data in column c1 will then have a mean and standard deviation that exactly matches that of the first subgroup. Repeat this procedure (use Minitab's Command Editor) for creating the remaining subgroups and store them in columns c2 through c24.*

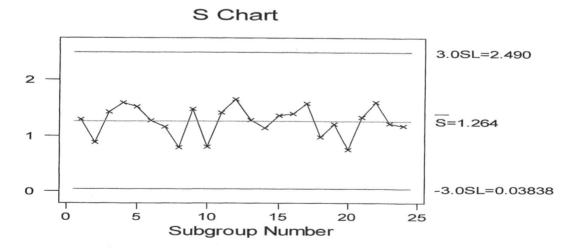

(b) A Minitab $\bar{x}$ chart of this data appears below. The centerline is 96.503. The control limits, based on the s chart centerline, are UCL $= \bar{\bar{x}} + A_3\bar{s} = 96.503+1.287(1.2642) =$ 98.1300 and LCL $= \bar{\bar{x}} - A_3\bar{s} = 96.503-1.287(1.2642) = 94.8760$.

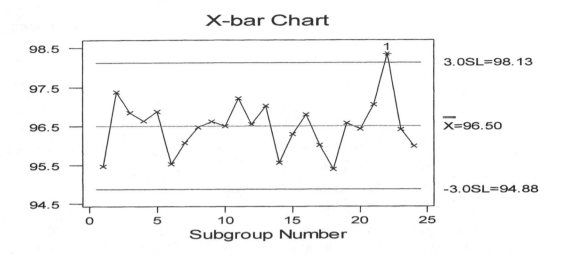

X-bar Chart

19.   (a) If each x_i value is transformed into $y_i = b(x_i - a)$, where a and b are constants and $b > 0$, then for any set of n values, $\bar{y} = b(\bar{x} - a)$ and $\bar{R}_y = b\bar{R}_x$. From these two relationships, simple algebra will show that, for example, $x_i > $ UCL (of the x-data) if and only if $y_i > $ UCL (of the y-data):
$$x_i > UCL_x = \bar{\bar{x}} + A_2\bar{R}_x \Rightarrow y_i = b(x_i\text{-}a) > b(\bar{\bar{x}}\text{-}a + A_2\bar{R}_x) = b(\bar{\bar{x}}\text{-}a) + A_2(b\bar{R}_x) = \bar{\bar{y}} + A_2\bar{R}_y$$
$= UCL_y$ That is, the x-bar charts based on un-transformed data and transformed data will give the same signals. In the same manner, it can be shown that the R charts for both transformed and un-transformed data give the same signals.

(b) The transformed data is shown below:

i	y_1	y_2	y_3	i	y_1	y_2	y_3
1	4	0	2	11	-1	3	0
2	-1	-3	-1	12	-2	-1	4
3	-2	4	2	13	4	-1	3
4	-2	-2	1	14	-3	3	2
5	0	-2	2	15	2	0	3
6	-1	0	2	16	-3	1	-1
7	-3	3	3	17	-2	2	1
8	-2	-3	1	18	-3	2	-1
9	-3	1	3	19	-1	-2	0
10	3	1	1	20	1	-2	-1

The centerline of the R chart is 4.100, $UCL = D_4 \overline{R} = (2.574)(4.10) = 10.5534$, $LCL = D_3 \overline{R} = (0)(4.10) = 0.0000$. The centerline of the x-bar chart is 0.2500, $UCL = \overline{\overline{x}} + A_2 \overline{R} = .25000 + (1.023)(4.10) = 4.444$, and $LCL = \overline{\overline{x}} - A_2 \overline{R} = .25000 - (1.023)(4.10) = -3.944$. There are no 'out of control' conditions in either chart.

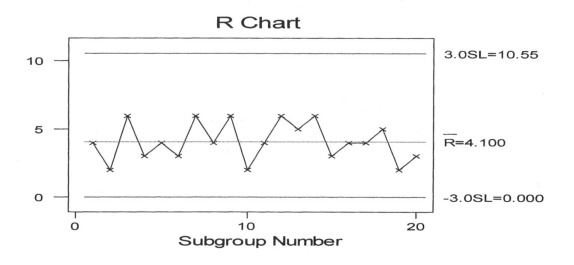

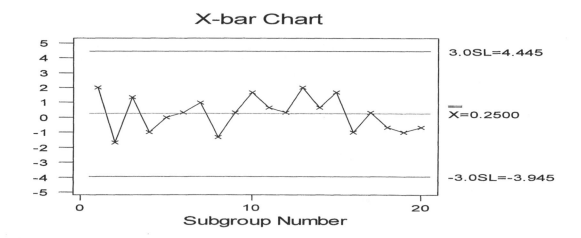

(c) The centerline of the R chart is 0.0041, UCL = $D_4 \bar{R}$ = (2.574)(.0041) = .010553, LCL = $D_3 \bar{R}$ = (0)(.0041) = 0.0000. The centerline of the x-bar chart is 0.2542 , UCL = $\bar{\bar{x}}$ +$A_2 \bar{R}$ = .2542+(1.023)(.0041) =.2584, and LCL = $\bar{\bar{x}}$ -$A_2 \bar{R}$ = .2542-(1.023)(.0041) = 0.2501. There are no 'out of control' conditions in either chart.

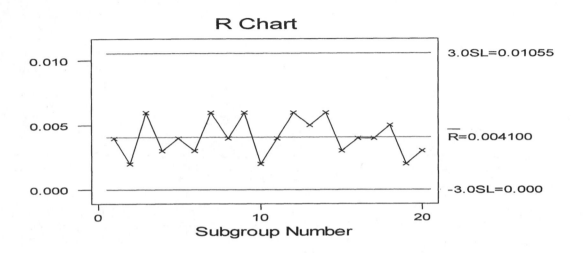

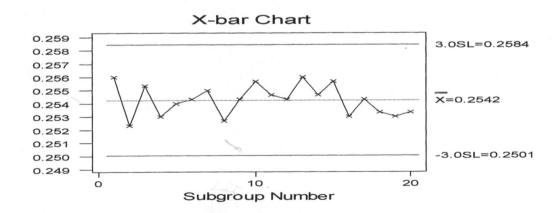

21.   (a) $\sigma \approx \frac{s}{c_4}$ = 0.043714/0.8862 = 0.04933.

(b) P(x > USL) = P(x > .48) $\approx$ P(z > (.48-.3813)/.04933) = P(z > 2.00) = .0028.
    P(x < LSL) = P(x < .32) $\approx$ P (z < (.32-.3813)/.04933) = P(z < -1.24) = .1075.

84

23.   The capability ratio $= 1/C_p = 1/1.2 = .833$. That is, 83.3% of the specification range is used by the process measurements.

25.   The magnitude of C_{pk} does <u>not</u> indicate which direction the mean has shifted, so the process average could be *either* of 5.0017 or 4.9983. These values are found as follows:

     USL $= 5+.01 = 5.01$ and LSL $= 5-.01 = 4.99$, so USL-LSL $= .02$. Therefore, $C_p = 1.2 =$ (USL-LSL)$/6\hat{\sigma} = .02/6\hat{\sigma}$, so $\hat{\sigma} = .02/(6(1.2)) = .002777$. To obtain a $C_{pk} = 1.0$, we must have *either* $\overline{x}$-LSL $= 3\hat{\sigma}$ *or* USL-$\overline{x} = 3\hat{\sigma}$. The first equation yields $\overline{x} =$ LSL$+3\hat{\sigma} =$ $4.99+3(.002777) = 4.9983$; the second yields $\overline{x} =$ USL-$3\hat{\sigma} = 5.01-3(.002777) = 5.0017$.

27.   Proportion out of spec $= P(z \geq 3C_{pk}) + P(z \geq 6C_p-3C_{pk}) = P(z \geq 3(1.0)) + P(z \geq 6(1.2)-3(1.0))$ $= P(z \geq 3.00) + P(z \geq 4.20) = 0.0013 + 0.0000 = 0.0013$, or, about .13%.

29.   From Problems 15 and 20, $\overline{x} = 0.38133$ and $\sigma \approx 0.04933$. The upper and lower specification limits are USL $= .40+.08 = .48$ and LSL $= .40-.08 = .32$, so USL - LSL $= .48-.32 = .16$. Therefore, $C_p =$ (USL-LSL)$/6\hat{\sigma} = .16/[(6)(.04933)] = .5406$. The k factor is
$$\frac{|(USL+LSL)/2-\hat{\mu}|}{(USL-LSL)/2} = \frac{|(.48+.32)/2-.38133|}{(.48-.32)/2} = .23338, \text{ so } C_{pk} = (1-k)C_p =$$
$(1-.23338)(.5406) = .4144$.

     Note: when the nominal value is midway between the specification limits, which is often the case, then USL-LSL can be found by simply doubling the plus/minus tolerance; e.g. in this problem, USL-LSL = 2(.08) = .16. Also, the nominal value equals (USL+LSL)/2, which simplifies the calculation of the k factor.

31.   Points on a p chart that are above the UCL are signs of high defect rates and such problems must be <u>eliminated</u>. However, points below the LCL are evidence of unusually *low* defect rates, which is a good thing, so the conditions that gave rise to such points should be investigated for their potential to make permanent process <u>improvements</u>.

33.   When the LCL is positive, $\overline{c} - 3\sqrt{\overline{c}} > 0$, or, $\overline{c} > 3\sqrt{\overline{c}}$. Dividing through by $\overline{c}$ yields the inequality $\sqrt{\overline{c}} > 3$. Squaring both sides, we find $\overline{c} > 9$. Conversely, starting with $\overline{c} > 9$, you can reverse the steps to show that LCL > 0.

35. (a) The centerline can be found from the formula $\bar{p} = \dfrac{x_1 + x_2 + \ldots + x_k}{n_1 + n_2 + \ldots + n_k} = \dfrac{578}{100(25)} =$

.2312. The upper and lower control limits are then $\bar{p} \pm 3\sqrt{\dfrac{\bar{p}(1-\bar{p})}{n}}$, or, UCL = .2312

$+ 3\sqrt{\dfrac{.2312(1-.2312)}{100}} = .2312 + 3(.04216) = 0.3577$ and LCL = .2312 $-$

$3\sqrt{\dfrac{.2312(1-.2312)}{100}} = .2312 - 3(.04216) = 0.1047.$

(b) These defect numbers translate into proportions of p = 39/100 = .39 and p = 13/100 = .13. The first of these two points, indicates and 'out-of-control' condition because p = .39 exceeds the UCL = .3577

(c) To eliminating the subgroup with p = 39 we subtract its number of defects from the total (i.e., 578-39 = 539) and subtract the subgroup size from the total of the subgroups (i.e., 100(25) - 100 = 2400). So, the new value of $\bar{p}$ is 539/2400 = .2246. The revised control limits are then: UCL = .2246 + 3(.04173) = .3498 and LCL = .2246 - 3(.04173) = .0994.

37. (a) A Minitab c chart of this data appears below. The centerline of the chart is $\bar{c} = 4.08$ and the control limits are UCL = $\bar{c} + 3\sqrt{\bar{c}} = 4.08 + 3(2.0199) = 10.1397$ and LCL = 0 (since $\bar{c} - 3\sqrt{\bar{c}} = 4.08 - 3(2.0199) = -1.9797$, a negative number).

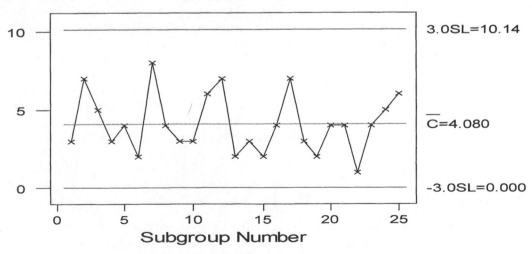

(b) There are no points are outside these control limits.

39.   (a) A Minitab c chart of this data appears below. The centerline of the chart is $\bar{c} = 3.925$ and the control limits are UCL = $\bar{c} + 3\sqrt{\bar{c}} = 3.925\ 3(1.98116) = 9.8685$ and LCL = 0 (since $\bar{c} - 3\sqrt{\bar{c}} = 3.925 - 3(1.98116) = -12.0185$, a negative number).

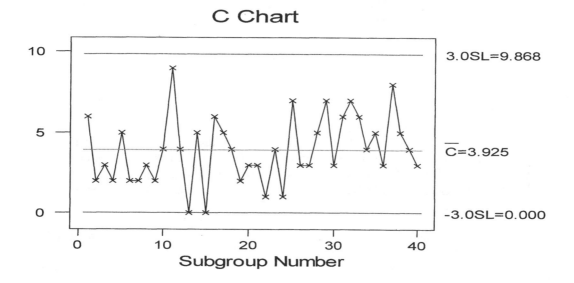

(b) There are no 'out-of control' conditions indicated on the chart, although two of the points do coincide with the lower control limit and could be worth investigating.

41.   (a) Let T denote the number of cycles that a given canula system will last. Assuming that T has an exponential distribution with $\mu = 500$, the density function is $f(t) = \lambda e^{-\lambda t}$. In this example, the parameter $\lambda = 1/\mu = 1/500 = .002$.

Recall (see the solution to Problem 23 in Chapter 1), that $P(T > c) = e^{-\lambda c}$ for any positive constant c. Therefore, the probability such a system lasts for at least 100 cycles is:

$$P(T > 100) = e^{-\lambda(100)} = e^{-.002\,(100)} = e^{-.2} = .8187.$$

(b) Set $R(t) = P(T > t) = .95$ and solve for t. From part(a), $P(T>t) = e^{-\lambda t}$, so the equation to solve becomes $e^{-\lambda t} = .95$. Taking natural logs of both sides gives $\ln(e^{-\lambda t}) = \ln(.95)$ which becomes $-\lambda t = -.051293$. Solving for t, we have $t = -.051293/(-\lambda) = -.051293/(-.002) = 25.6466$. That is, the reliability will be about 95% at $t \approx 26$ cycles.

(c)   To have R(100) = .95, the mean time to failure, μ, must be increased. The new value of μ can be found by putting $t = 100$ into the formula $R(t) = e^{-\lambda t}$, setting the result equal to .95 and solving for λ (and hence μ). Thus, $e^{-\lambda(100)} = .95$, so by taking natural logs of both sides: $-\lambda(100) = \ln(.95) = -.051293$ and therefore $\lambda = -.051293/(-100) = .00051293$. Finally, $\mu = 1/\lambda = 1/.00051293 = 1949.58$. So, the mean time to failure would have to be increased from 500 cycles to about 1950 cycles in order to achieve a reliability of 95% at $t = 100$ cycles.

43.   (a)   Recall from problem 41 (or from problem 23 in Chapter 1) that for an exponentially distributed variable X with parameter λ, $P(X > t) = e^{-\lambda t}$. Since the mean μ of an exponential distribution is related to the parameter λ by $\lambda = 1/\mu$, the value of λ is $1/75 = .01333$ in this problem. If such an exponential distribution were applied to a human population, then the proportion of people living over 150 years would be $P(X > 150) = e^{-\lambda(150)} = e^{-150/75} = e^{-2} \approx .135$. This is clearly impossible, since it says that 13.5% of the population would live to be over 150 years old (and empirical data shows that no one has ever lived that long).

(b)   On the other end of the spectrum, the proportion of people living less than 10 years would be $P(X < 10) = 1 - P(X \geq 10) = 1 - e^{-10/75} = 1 - e^{-.13333} = 1 - 8752 \approx .125$. Again, this is unreasonable since it says that 12.5% of the population would die before age 10, a proportion far greater than what actually happens.

(c)   Obviously, the exponential distribution gives unreasonable estimates of the proportions of human populations that die before age 10, after age 150, and at all other ages, so it is <u>not</u> a good choice for modeling human lifetimes.

(d)   The "memoryless" property of the exponential distribution says that $P(X > t+s \mid X > s) = P(X > t)$. That is, if a person has already lived s years, then the probability they live at least another t years is the same as the probability that a person lives to at least age t. Suppose, for example, $s = 80$ and $t = 20$. Then this statement would say that it is just as likely for an 80 year-old person to live until 100 as it is for a newborn to live until at least age 20. This does not match our experience with human populations so, again, the exponential distribution would not be a good choice for modeling human lifetimes.

45.   (a)

i:	1	2	3	4	5	6	7	8	9	10
x_i:	.47	.58	.65	.69	.72	.74	.77	.79	.80	.81
p_i:	.01278	.0833	.1389	.1944	.2500	.3056	.3611	.4167	.4722	.5278

i:	11	12	13	14	15	16	17	18
x_i:	.82	.84	.86	.89	.91	.95	1.01	.104
p_i:	.5833	.6389	.6944	.7500	.8056	.8611	.9167	.9722

A regression analysis of $\ln(\ln(\frac{1}{1-p_i}))$ on $\ln(x_i)$ yields the estimated model $6.308758\ln(x)$ + .979962. Thus, $\alpha = 6.308758$ and $-\alpha\ln(\beta) = .979962$, which means that $\ln(\beta) = -.979962/\alpha$ = $-.979962/6.308758 = -.1553336$ and therefore, $\beta = .8561295$.

(b).

parameter	Regression estimates	Maximum Likelihood estimates (page 332 of the text)
α	6.309	6.653
β	.856	.854

47.   Adhesive strength has a *two-sided tolerance* since the strength can not be too large or too small; i.e., it is required to be bounded above and below..

49.   The within-subgroups variation is completely lost using this method. Without a measure of the within-subgroups variation, there is no way to tell if the variation amongst the subgroup means is large or small. There will be very few, if any, subgroup means that fall out side the control limits based on $\overline{\overline{x}} \pm 3s^*$.

51.   (a) A Minitab s chart of this data appears below. The centerline of the chart is $\overline{s} = 0.20396$ and the control limits are UCL $= B_4\overline{s} = (1.815)(.20396) = 0.37019$, LCL $= B_3\overline{s} = (.185)(.20396) = 0.03773$.

(b) The point 0.388 is above the UCL; eliminating this point gives 23 points with $\overline{s} = 0.1960$ and associated control limits UCL $= 0.3557$, LCL $= 0.0363$. No points are outside these control limits.

53.   Refer to Problem 6 in this chapter. Remember that the estimate of the out-of-spec proportion applies to the entire process, not to the particular set of data in the sample.

55.   First, $\overline{p} \neq 0$ since that would imply that LCL $= 0$. So, for a given value of $\overline{p} > 0$, LCL is positive when LCL $= \overline{p} - 3\sqrt{\frac{\overline{p}(1-\overline{p})}{n}} > 0$, which means that $\overline{p} > 3\sqrt{\frac{\overline{p}(1-\overline{p})}{n}}$. Dividing through by $\overline{p}$ gives $1 > 3\sqrt{\frac{1(1/\overline{p}-1)}{n}}$ and squaring both sides gives $1 > 9(\frac{1/\overline{p}-1}{n})$, so $n/9 > 1/\overline{p} - 1$ and therefore, $n/9+1 > 1/\overline{p}$. Taking reciprocals of both sides, $\overline{p} > 1/(n/9+1) = 9/(n+9)$. The steps can also be reversed to show that $\overline{p} > 9/(n+9)$ implies LCL > 0.

57.   The key to the proof is to realize that the function $x(1-x) \le \frac{1}{4}$ for all values of x. Simple calculus shows this to be true. Suppose, for a given value of t that, say, $R_1(t)$ is the smaller of $R_1(t)$ and $R_2(t)$, so that $R_1(t) = \min[R_1(t), R_2(t)]$. Next, substitute $R_2(t)$ for x in the inequality above to find that $R_2(t) \cdot [1-R_2(t)] \le \frac{1}{4}$. Now, since $R_1(t)$ is smaller than $R_2(t)$, we must have $R_1(t) \cdot [1-R_2(t)] \le R_2(t) \cdot [1-R_2(t)] \le \frac{1}{4}$. That is, $R_1(t) \cdot [1-R_2(t)] \le \frac{1}{4}$, which can be rearranged to give $R_1(t) \le R_1(t) \cdot R_2(t) + \frac{1}{4}$. Since $R_1(t)$ is smaller than $R_2(t)$, we have $\min[R_1(t), R_2(t)] = R_1(t) \le R_1(t) \cdot R_2(t) + \frac{1}{4}$, which shows that $\min[R_1(t), R_2(t)] \le R_1(t) \cdot R_2(t) + \frac{1}{4}$ for the given value of t. If, on the other hand, $R_2(t)$ is the smaller of the two for some other value of t, then the same type argument can be applied to again show that $\min[R_1(t), R_2(t)] \le R_1(t) \cdot R_2(t) + \frac{1}{4}$.

59.   Components 1 and 2 are connected in series, so the reliability of that part of the system is $R_1(t) \cdot R_2(t)$. Next, because 1 & 2 are connected in parallel with 3, the system reliability R(t) is given by:

$$R(t) = 1 - [1-R_3(t)] \cdot [1- R_1(t) \cdot R_2(t)]$$

Chapter 7
Estimation and Statistical Intervals

1.   A single randomly selected item forms a sample of size n = 1. If x denotes the value of this item, then the sample average is also equal to x. Since the sample average, based on any sample size, is an unbiased estimator or the population mean, then the length x is indeed an unbiased estimator of the population mean. Of course, this estimator is not as precise as the sample average based on larger sample sizes (i.e., it has a larger standard error), but it is still unbiased.

3.   (a) Because the population is known to be normal and its standard deviation is known to be $\sigma = 5$, the sampling distribution of $\bar{x}$ is also normal. Standardizing, $P(\mu\text{-}1 \leq \bar{x} \leq \mu + 1)$

$$= P(\frac{(\mu-1)-\mu}{\sigma/\sqrt{n}} \leq z \leq \frac{(\mu+1)-\mu}{\sigma/\sqrt{n}}) = P(-\frac{\sqrt{n}}{\sigma} \leq z \leq \frac{\sqrt{n}}{\sigma}) = P(-.63 \leq z \leq .63) =$$

.7357 - .2643 =   .4714.

(b) In each case, as in part (a), the desired probability is $P(-\frac{\sqrt{n}}{\sigma} \leq z \leq \frac{\sqrt{n}}{\sigma})$:

For n = 50, $P(-\frac{\sqrt{n}}{\sigma} \leq z \leq \frac{\sqrt{n}}{\sigma}) = P(-1.41 \leq z \leq 1.41) = .9207 - .0793 = .8414.$

For n = 100, $P(-\frac{\sqrt{n}}{\sigma} \leq z \leq \frac{\sqrt{n}}{\sigma}) = P(-2.00 \leq z \leq 2.00) = .9772 - .0228 = .9544.$

For n = 1000, $P(-\frac{\sqrt{n}}{\sigma} \leq z \leq \frac{\sqrt{n}}{\sigma}) = P(-6.32 \leq z \leq 6.32) \approx 1.0000 - .0000 = 1.0000.$

In other words, as n increases, it becomes more and more likely that the sample mean will fall within ±1 standard deviation from the population mean.

(c) As the following graph shows, the probability that the sample mean lies within ±1 unit of μ increases as the sample size n increases:

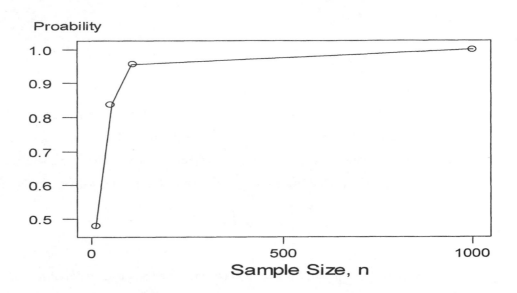

5.   (a) As in exercise 3, we have $P(\mu-1 \le \bar{x} \le \mu + 1) = P(\dfrac{(\mu-1)-\mu}{\sigma/\sqrt{n}} \le z \le \dfrac{(\mu+1)-\mu}{\sigma/\sqrt{n}})$

$= P(-\dfrac{\sqrt{n}}{\sigma} \le z \le \dfrac{\sqrt{n}}{\sigma})$. In this exercise, $\sigma = 2$. To make this probability equal to 90%, first use Table I (page 561) to find the (approximate) value $z = 1.645$ for which $P(-1.645 \le z \le 1.645) \approx .90$. Then equate 1.645 and $\dfrac{\sqrt{n}}{\sigma}$ and solve for n: $\dfrac{\sqrt{n}}{\sigma} = 1.645$, so $\sqrt{n} = 1.645(\sigma) = 1.645(2)$ and, squaring, $n = (4)(1.645)^2 = 10.82$. Rounding off to the nearest integer value, $n \ge 11$ will guarantee that there is at least a 90% probability that the sample mean (based on a sample of size 11) will fall within ±1 standard deviation of the population mean.

(b) Using Table I (pages 534-535): $P(-1.28 \le z \le 1.28) \approx 80\%$. Therefore, $\dfrac{\sqrt{n}}{\sigma} = 1.28$, so $\sqrt{n} = 1.28(\sigma) = 1.28(2)$ and $n = 4(1.28)^2 = 6.55$, so $n \ge 7$.

Similarly, $P(-1.96 \le z \le 1.26) \approx 95\%$. Therefore, $\dfrac{\sqrt{n}}{\sigma} = 1.96$, so $\sqrt{n} = 1.96(\sigma) = 1.96(2)$ and $n = 4(1.96)^2 = 15.4$, so $n \ge 16$. *Note: in every case we round up to the next higher integer because we want to be sure of having exceeding the specified probability requirement.*

92

Finally, $P(-2.575 \le z \le 2.575) \approx 99\%$. Therefore, $\dfrac{\sqrt{n}}{\sigma} = 2.575$, so $\sqrt{n} = 2.575(\sigma) = 2.575(2)$

and $n = 4(2.575)^2 = 26.52$, so $n \ge 27$.

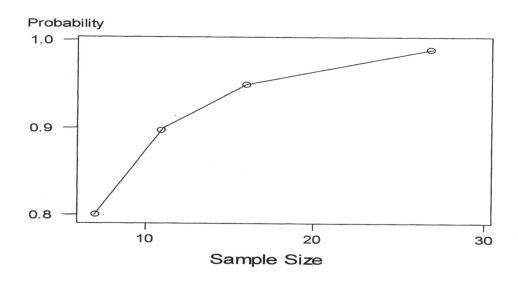

7.   (a) The entry in the 3.0 row and .09 column of the z table is .9990. Similarly, the entry for
           -3.09 is .0010. Therefore, the area under the z curve between -3.09 and +3.09 is .9990
           - .0010 = .9980. The confidence level is then 99.8%.

      (b) Following the example in part (a), the z-table entries corresponding to z = -2.81 and
           z = +2.81 are .9975 and .0025, respectively. Therefore the area between these two z
           values is .9975 - .0025 = .9950. The confidence level is then 99.5%.

      (c) The z-table entries corresponding to z = -1.4 and z = +1.44 are .9251 and .0749,
           respectively. Therefore the area between these two z values is .9251 - .0749 = ..8502.
           The confidence level is then 85.02%.

      (d) The coefficient of s/$\sqrt{n}$ is not written, but is understood to be 1.00. The z-table entries
           corresponding to z = -1.001 and z = +1.00are .8413 and .1587, respectively.
           Therefore the area between these two z values is .8413 - .1587 = .6826. The confidence
           level is then 68.26%.

9.   (a) The larger the confidence level, the larger the z value. Since the plus-or-minus width of the confidence interval equals $z\frac{s}{\sqrt{n}}$, the <u>larger z values will result in wider confidence intervals</u>. (Conversely, smaller confidence levels will result in narrower confidence intervals.)

    (b) Because n appears in the denominator of the expression $z\frac{s}{\sqrt{n}}$, <u>larger values of n will result in narrower confidence intervals</u>. (Conversely, smaller sample sizes will result in wider confidence intervals.)

    (c) Because s appears in the numerator of the expression $z\frac{s}{\sqrt{n}}$, <u>larger values of s will result in wider confidence intervals</u>. (Conversely, smaller values of s will result in narrower confidence intervals.)

11.   (a) Decreasing the confidence level from 95% to 90% will decrease the associated z value and therefore make the 90% interval narrower than the 95% interval. *(Note: see the answer to Exercise 9 above)*

    (b) The statement is not correct. One a particular confidence interval has been created/calculated, then the true mean is either in the interval or not. The 95% refers to the process of creating confidence intervals; i.e., it means that 95% of all the possible confidence intervals you could create (each based on a new random sample of size n) will contain the population mean (and 5% will not).

    (c) The statement is not correct. A confidence interval states where plausible values of the population mean are, not where the individual data values lie. In statistical inference, there are three types of intervals: **confidence intervals** (which estimate where a population mean is), **prediction intervals** (which estimate where a single value in a population is likely to be), and **tolerance intervals** (which estimate the likely range of values of the items in a population. The statement in this exercise refers to the likely range of all the values in the population, so it is referring to a tolerance interval, not a confidence interval.

(d) No the statement is not exactly correct, but it is close. We *expect* 95% of the intervals we construct to contain μ, but we also expect a little variation. That is, in any group of 100 samples, it is possible to find only, say, 92 that contain μ. In another group of 100 samples, we might find 97 that contain μ, and so forth. So, the 95% refers to the *long run* percentage of intervals that will contain the mean. 100 samples/intervals is <u>not</u> the long run.

13. The sample size (n = 110) is large enough to allow you to use the 'large-sample' confidence interval formula, $\bar{x} \pm z\frac{s}{\sqrt{n}}$. The z-critical value for a 99% confidence interval is approximately 2.58. To find z = 2.58, note that to obtain an area of .99 symmetrically located in the middle of the z distribution, the right- and left-tail areas must each equal .005. In the z table, the value of z = -2.57 corresponds to a left-tail area of .0051, while z = -2.58 corresponds to an area of .0049. Interpolating would give z = -2.575 (a more precise value is 2.576, which is sometimes given in other texts), but we will just round to two places and use -2.58. Therefore, the 99% confidence interval for μ is $\bar{x} \pm z\frac{s}{\sqrt{n}}$ = .81 ± (2.58) $\frac{.34}{\sqrt{110}}$ = .81 ± .0836 = [.7264, .8936], or, about [.73, .89]. We can be highly confident that .73 < μ < .89.

15. (a) The large sample size (n = 169) allows us to use the large-sample formula $\bar{x} \pm z\frac{s}{\sqrt{n}}$.

The z-value for 95% confidence is z = 1.96, so the interval is $\bar{x} \pm z\frac{s}{\sqrt{n}}$ = 89.10 ± (1.96) $\frac{3.73}{\sqrt{169}}$ = 89.10 ± .5624 = [88.5376, 89.6624] or, approximately, [88.54, 89.66]. The interval is very narrow (i.e., its width 89.66-88.54 = 1.12 is a small fraction of the mean value, 89.10) so our knowledge about μ is quite precise.

(b) Given that σ = 4, the minimum sample size needed to estimate μ to within ±.5 Mpa is:

$$n = \left(\frac{1.96\sigma}{B}\right)^2 = \left(\frac{1.96(4)}{.5}\right)^2 = (15.68)^2 = 245.86, \text{ or, } n \approx 246.$$

17. (a) The entry in the z table corresponding to z = .84 is .7995, so the confidence level is 79.95% or, approximately, 80%.

(b) The entry in the z table corresponding to z = 2.05 is .9798, so the confidence level is 97.98% or, approximately, 98%.

(c) The entry in the z table corresponding to z = .67 is .7486, so the confidence level is 74.86% or, approximately, 75%.

19. The formula for a lower confidence bound is $\bar{x} - z\frac{s}{\sqrt{n}}$, where the z critical value corresponds to a left tail area of 99%. From the z table, z = 2.33 comes closest to meeting this requirement, so the 99% lower confidence bound is 64.3 - (2.33) $\frac{6.0}{\sqrt{32}}$ = 64.3 - 2.471 = 61.83.

21. n = 356 and there are x = 201 'successes', so the sample proportion of successes is p = 201/365 = .5646. The traditional (large n) confidence interval formula is $p \pm z\sqrt{\frac{p(1-p)}{n}}$, where z = 1.96 (for 95% confidence). Plugging in the sample value of p: .5646 ± (1.96) $\sqrt{\frac{.5646(1-.5646)}{356}}$ = .5646 ± .0515 = [.5131, .6161]. Therefore, in the long run, we expect the percentage of dies that pass the probe will be somewhere between 51.31% and 61.61%. This interval is not extremely narrow (e.g., the width is 61%-51% = 10%, which is about 10/56.46 = 17.7% of the average value), so this interval does not give great precision in estimating the true proportion π. Using the improved approximate confidence interval formula (cf. box on page 305 of the text), the confidence limits are [.5127, .6151], which are very close to those given by the traditional formula.

23. The sample proportion of households owning at least one gun is $p = 133/549 = .2467$. For a 95% one-sided confidence bound, we use a z value of 1.645 (which has a cumulative left-tail area of approximately .95). The bound is $p - z\sqrt{\dfrac{p(1-p)}{n}} =$

$.2467 - (1.645)\sqrt{\dfrac{.2467(1-.2467)}{549}} = .2467 - .0303 = .2164$, or about 21.6%. We estimate, with 95% confidence, that at least 21.6% of all households own at least one firearm. Using the improved approximate confidence interval formula (page 305 of the text), the confidence bound is ..217, which is very close to that given by the traditional formula.

25. (a) Following the same format used for most confidence intervals, i.e., *statistic ± (critical value)(standard error)*, an interval estimate for $\pi_1 - \pi_2$ is:

$$(p_1 - p_2) \pm z\sqrt{\dfrac{p_1(1-p_1)}{n_1} + \dfrac{p_2(1-p_2)}{n_2}}.$$

(b) The response rate for no-incentive sample: $p_1 = 75/110 = .6818$, while the return rate for the incentive sample is $p_2 = 66/98 = .6735$. Using $z = 1.96$ (for a confidence level of 95%), a two-sided confidence interval for the true (i.e., population) difference in response rates $\pi_1 - \pi_2$ is:

$$(.6818 - .6735) \pm (1.96)\sqrt{\dfrac{.6818(1-.6818)}{110} + \dfrac{(.6735)(1-.6735)}{98}} = .0083 \pm .1273 =$$

$[-.119, .1356]$. The fact that this interval contains 0 as a plausible value of $\pi_1 - \pi_2$ means that it is plausible that the two proportions are equal. Therefore, including incentives in the questionnaires does not appear to have a significant effect on the response rate.

(c) The new $p_1 = 76/112 = .67857$ and the new $p_2 = 67/100 = .67$. Using these values in the calculations above, the confidence interval is $.00857 \pm .1264 = [-.118, .135]$

27.   (a) As in Exercise 25, the usual confidence interval format

statistic ± (critical value)(standard error) gives a confidence interval for $\ln(\pi_1/\pi_2)$ of:

$$\ln(p_1/p_2) \pm z\sqrt{\frac{n_1-u}{n_1 u} + \frac{n_2-v}{n_2 v}} \ .$$

(b) Since we want to estimate the ratio of return rates for incentive group to the non-incentive group, we will call group 1 the incentive group (to match the subscripts in the formula above). The number of returns for the non-incentive group is $v = 75$ out of $n_2 = 110$, so $p_2 = .6818$. For the incentive group, $u = 78$ out of $n_1 = 100$, so $p_1 = .78$. The 95% confidence interval for $\ln(\pi_1/\pi_2)$ is:

$$\ln(.7800/.6818) \pm (1.96)\sqrt{\frac{100-78}{100(78)} + \frac{110-75}{110(75)}} = .1346 \pm .1647 = [-.0301, .2993]. \text{ To find}$$

a 95% interval for π_1/π_2 , we exponentiate both endpoints of this interval, $[e^{-.0301}, e^{.2993}] =$ [.9702, 1.3489], or, about [.970, 1.349]. Since this interval includes the 1, it is plausible that $\pi_1/\pi_2 = 1$; i.e., it is plausible that the return rates are equal. As in problem 25, the use of the incentive does not appear to have an effect on the questionnaire return rate.

29.   For 90% confidence, the associated z value is 1.645. Since nothing is known about the likely values of π we use .25, the largest possible value of $\pi(1-\pi)$, in the sample size formula:

$$n = (.25)\left(\frac{1.645}{.05}\right)^2 = 270.6. \text{ To be conservative, we round this value up to the next highest}$$

integer and use $n = 271$.

31.   Let μ_1 denote the average density of the population having a low percentage of juvenile wood and let μ_2 denote the average density for the population with a moderate percentage of juvenile wood. To estimate the difference between the two means with, say, 95%

confidence, we use the large-sample formula: $(\bar{x}_1 - \bar{x}_2) \pm z\sqrt{\frac{s_1^2}{n_1} + \frac{s_2^2}{n_2}} = (.523 - .489) \pm$

$(1.96)\sqrt{\frac{(.0543)^2}{35} + \frac{(.0450)^2}{54}} = .034 \pm .0216 = [.0124, .0556].$ The interval is not too

narrow compared to the difference between the average values of the densities, but it does indicate that there seems to be a slight positive difference between the population means (since both endpoints of the interval are positive, indicating that $\mu_1-\mu_2$ is most likely positive). That is, we can conclude that the average density for the 'low %' population is probably larger than the average density for the 'moderate %' population.

33.   Let μ_1 denote the average toughness for the high-purity steel and let μ_2 denote the average toughness for the commercial purity steel. Then, a lower 95% confidence bound for $\mu_1-\mu_2$ is

given by: $(\bar{x}_1 - \bar{x}_2) - z\sqrt{\dfrac{s_1^2}{n_1} + \dfrac{s_2^2}{n_2}} = (65.6-59.2) - (1.645)\sqrt{\dfrac{(1.4)^2}{32} + \dfrac{(1.1)^2}{32}} = 6.4 - .518 =$

5.882. Because this lower interval bound exceeds 5, it gives a reliable indication that the difference between the population toughness levels does exceed 5.

35.   (a) From Table IV (row 10, central area = .95), the critical value is 2.228.

       (b) From Table IV (row 20, central area = .95), the critical value is 2.086.

       (c) From Table IV (row 20, central area = .99), the critical value is 2.845.

       (d) A table entry for df = 50 is not given in Table IV, so we interpolate between the table values for df =40 and df = 60: the value is approximately (2.704 + 2.660)/2 = 2.682

       (e) An upper tail area of .01 corresponds to a cumulative area of .99, so from Table IV (row 25, cumulative area area = .99), the critical value is 2.485.

       (f) A lower tail area of .025 corresponds to a central area of .95, so from Table IV (row 5, central area = .95), the critical value is 2.571.

37.   In the following answers, we give the critical values are found by selecting the column in Table IV corresponding to the desired cumulative area (which equals the confidence level):

(a) Cumulative area = .95, df = 10: t critical value = 1.812.

(b) Cumulative area = .95, df = 15: t critical value = 1.753.

(c) Cumulative area = .99, df = 15: t critical value = 2.602.

(d) Cumulative area = .99, df = n-1 = 5-1 =4: t critical value = 3.747.

(e) There is no 98% cumulative area column, so we interpolate between the values in the nearby 97.5% and 99% columns: for 24 df, these values are 2.064 and 2.492. The 98% value is *about* 2.064 + (.98-.975)/(.99-.975)[2.492-2.064] = 2.064 + (.3333)[.428] = 2.207.

(f) Since there is no entry for df = n-1 = 38-1 = 37, we must interpolate between the entries for df = 30 and df = 40: critical value ≈ 2.457 + (37-30)/(40-30)[2.423 - 2.457] = 2.4332 or, about 2.43.

39.   Recall that this text's definition of upper and lower quartiles may differ a little from the definitions used by some other authors. We define the upper and lower quartiles to be the medians of the lower and upper halves of the data (for n odd, the median is included in both halves). This usually gives results that are very close to the actual (or estimated) quartiles used by other authors, but there are often small differences. For example, in this exercise the medians of the lower and upper halves of the data are: lower quartile = median of lower half (including the median, 437) = 425; upper quartile = median of the upper half (including median) = 448. The software package Minitab, however, estimates the lower quartile as the $.25(n+1)^{th} = .25(17+1)^{th} = 4.5^{th}$ item in the sorted list (here, 422 and 425 are the 4th and 5th items in the sorted list, and the 4.5^{th} value is defined to be their average, which is 423.5). So, be careful when comparing your answers to those in various software packages.

(a) Using 425 and 448 as the lower and upper quartiles, the IQR = 448 - 425 = 23. To check for outliers, we calculate the values 425 + 1.5(IQR) = 425 - 1.5(23) = 390.5 and 448 + 1.5(IQR) = 448 - 1.5(23) = 482.5. Since the maximum and minimum in the data are 465 and 418, which are inside the limits just calculated, we conclude that there are no outliers in the data. The median of the data is 437 and a boxplot of the data appears below:

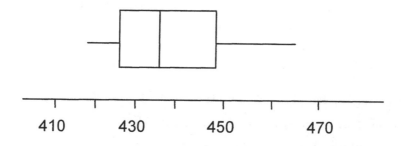

(b) A quantile plot can be used to check for normality. Refer to the answer to Exercise 43 of Chapter 2 for an easy method of creating a quantile plot in Minitab. The quantilile plot below shows a fairly strong linear pattern, supporting the assumption that this data came from a normal population:

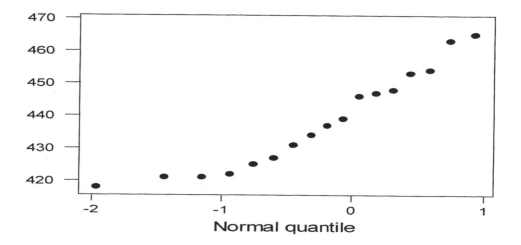

(c) Since the sample size is small (n = 17), we use an interval based on the t distribution. The sample mean and standard deviation are 438.29 and 15.144, respectively. For n -1 = 17 -1 = 16 df, the critical t value for a 2-sided 95% confidence interval is 2.120 (from Table IV). Therefore, the desired confidence interval is $438.29 \pm (2.120)\,^{15.144}/\sqrt{17}$ = $438.29 \pm 7.787 = [430.5, 446.1]$. This interval suggests that 440 (which is inside the interval) is a plausible value for the mean polymerization. The value of 450, however, is not plausible, since it lies outside the interval.

41. (a) The sample mean and standard deviation of the data are 2887.6 and 84.026, respectively. Because the sample size is small (n = 5), a confidence interval based on the t distribution is appropriate. For n -1 = 5 -1 = 4 df, the critical t value for a 2-sided, say 95%, confidence interval is 2.776 (from Table IV). The desired confidence interval is $2887.6 \pm (2.776)\,^{84.026}/\sqrt{5} = 2887.6 \pm 104.32 = [2783.28, 2991.92]$.

(b) A 95% prediction interval for a single asphalt specimen is: $\bar{x} \pm t \cdot s \sqrt{1 + \frac{1}{n}} = 2887.6 \pm$ $(2.776)(84.026)\sqrt{1 + \frac{1}{5}} = 2887.6 \pm 255.52 = [2632.1, 3143.1]$. The prediction interval is much wider than the confidence interval in part (a), which is expected since prediction intervals are always wider than confidence intervals (for a given confidence level).

43. (a) The sample mean and standard deviation of the data are .9255 and .0809, respectively. Because the sample size (n = 20) is small, an interval based on the t distribution is appropriate. The critical t value associated with df = n-1 - 19 is 2.093 (Table IV), so the desired interval is: $.9255 \pm (2.093)\,^{.0809}/\sqrt{20} = .9255 \pm .0379 = [.888, .963]$. Therefore, we can be reasonably confident that $.888 < \mu < .963$.

(b) A 95% prediction interval based on the same data would use the same critical t value:

$$\bar{x} \pm t \cdot s \sqrt{1 + \tfrac{1}{n}} = .9255 \pm (2.093)(.0809) \sqrt{1 + \tfrac{1}{20}} = .9255 \pm .1735 = [.752, 1.099].$$

Therefore, we can be reasonably confident that a single individual selected from the population will have a cadence between about .75 and 1.10.

(c) This question asks for a 99% tolerance interval. Appendix Table V gives the required critical value of 3.615 for a 99% tolerance interval based on a sample of size 20. The tolerance interval is then: $\bar{x} \pm ts = .9255 \pm 3.615(.0809) = .9255 \pm .2925 = [.633, 1.218]$.

45. (a) Confidence level = area between -.687 and 1.725 = .95 - .25 = .70, or, 70%. Note that the value 1.725 is in Table IV and is associated with a cumulative area of .95 (for df = 20).

(b) Confidence level = area between -.860 and 1.325 = .90 - .20 = .70, or, 70%. Note that the value 1.325 is in Table IV and is associated with a cumulative area of .90 (for df = 20).

(c) Confidence level = area between -1.064 and 1.064 = (1-2(.15)) = .70, or, 70%. Note that the symmetry of the t distribution allows us to conclude that the area to the left of 1.064 is also .15.

All three intervals have 70% confidence, but they have different widths. The interval in part (c) has the smallest width of $2(1.064) \, {s}/{\sqrt{n}}$, and is therefore the best choice among the three.

47. The approximate degrees of freedom for this estimate are: $df \approx \dfrac{\left[\frac{5.5^2}{28} + \frac{7.8^2}{31} \right]^2}{\dfrac{\left[\frac{5.5^2}{28} \right]^2}{27} + \dfrac{\left[\frac{7.8^2}{31} \right]^2}{30}} =$

53.954, or, about 54 (normally we would round down to 53, but this number is very close to 54 - of course for this large number of df, using either 53 or 54 won't make much difference in the critical t value). For a 90% 2-sided confidence interval with 54 df, the critical t value is about 1.68 (interpolate in Table IV), so the desired interval is: $(\bar{x}_1 - \bar{x}_2) - t \sqrt{\dfrac{s_1^2}{n_1} + \dfrac{s_2^2}{n_2}} =$

$$(91.5 - 88.3) \pm (1.68) \sqrt{\frac{5.5^2}{28} + \frac{7.8^2}{31}} = 3.2 \pm (1.68)(1.7444) = 3.2 \pm 2.931 = [.269, 6.131].$$

Because the value 0 does not lie inside this interval, we can be reasonably certain that the true difference $\mu_1-\mu_2$ is not 0 and, therefore, that the two population means are not equal. For a 95% interval, the t value increases to about 2.01 or so, which results in the interval 3.2 ± 3.506. Since this interval does contain 0, we can no longer conclude that the means are different if we use a 95% confidence interval.

49.   The approximate degrees of freedom for this estimate are: $df \approx \dfrac{\left[\frac{11.3^2}{6} + \frac{8.3^2}{8}\right]^2}{\left[\frac{11.3^2}{6}\right]^2 \Big/ 5 + \left[\frac{8.3^2}{8}\right]^2 \Big/ 7}$

$= 893.586/101.175 = 8.83$, so we round down and use $df \approx 8$. For a 95% 2-sided confidence interval with 8 df, the critical t value is 2.306, so the desired interval is: $(\bar{x}_1 - \bar{x}_2) -$

$$t \sqrt{\frac{s_1^2}{n_1} + \frac{s_2^2}{n_2}} = (40.3-21.4) \pm (2.306) \sqrt{\frac{11.3^2}{6} + \frac{8.3^2}{8}} = 18.9 \pm (2.306)(5.4674) = 18.9 \pm$$

$12.607 = [6.3, 31.5]$. Because 0 is not contained in this interval, there is strong evidence that $\mu_1-\mu_2 \neq 0$; i.e., we can conclude that the population means are not equal. Calculating a confidence interval for $\mu_2-\mu_1$ would change only the order of subtraction of the sample means (i.e., $\bar{x}_2 - \bar{x}_1 = -18.9$), but the standard error calculation would give the same result as before. Therefore, the 95% interval estimate of $\mu_2-\mu_1$ would be $[-31.5, -6.3]$, just the negatives of the endpoints of the interval for $\mu_1-\mu_2$. Since 0 is not in this interval, we reach exactly the same conclusion as before: the population means are not equal.

51.   (a) Following the usual format for most confidence intervals:
statistic $\pm$ (critical value)(standard error), a pooled-variance confidence interval for the difference between two means is: $(\bar{x}_1 - \bar{x}_2) \pm t \cdot s_p \sqrt{\frac{1}{n_1} + \frac{1}{n_2}}$, where t is the critical t value based on $df = n_1+n_2-2$.

(b) The sample means and standard deviations of the two samples are $\bar{x}_1 = 13.90$, $s_1 = 1.225$, $\bar{x}_2 = 12.20$, $s_2 = 1.010$. The pooled variance estimate is $s_p^2 =$

$$\frac{(n_1-1)s_1^2+(n_2-1)s_2^2}{n_1+n_2-2}=\frac{(4-1)(1.225)^2+(4-1)(1.010)^2}{4+4-2}=1.260, \text{ so } s_p=1.1227.$$

Since df = 4+4-2 = 6 for this interval, the critical t value for a 95% interval is 2.447 (from Table IV). The desired confidence interval is then: (13.90 - 12.20) ±

(2.447)(1.1227) $\sqrt{\frac{1}{4}+\frac{1}{4}}$ = 1.7 ± 1.943 = [-.24, 3.64]. This interval contains 0, so it does <u>not</u> support the conclusion that the two population means are different.

(c) The approximate degrees of freedom for this estimate are: df ≈ $\dfrac{\left[\frac{1.225^2}{4}+\frac{1.010^2}{4}\right]^2}{\left[\frac{1.225^2}{4}\right]^2\Big/3+\left[\frac{1.010^2}{4}\right]^2\Big/3}$

= 5.79 ≈ 5. The critical t value for df = 5 and 95% confidence is 2.571. *Note: the approximate df formula and the formula for pooled variances (df = n₁+n₂-2) are not necessarily equal, but are often close.* The desired confidence interval is then:

(13.90 - 12.20) ± (2.571) $\sqrt{\dfrac{1.225^2}{4}+\dfrac{1.010^2}{4}}$ = 1.7 ± 2.041 = [-.34, 3.74]. This interval

is a tiny bit wider than the interval in part (b) and gives the same conclusion regarding the population means: i.e., it does not suggest that the population means differ.

53. (a) The median of the 'Normal' data is 46.80 and the upper an lower quartiles are 45.55 and 49.55, which yields an IQR of 49.55 - 45.55 = 4.00. The median of the 'High' data is 90.1 and the upper an lower quartiles are 88.55 and 90.95, which yields an IQR of 90.95 - 88.55 = 2.40.

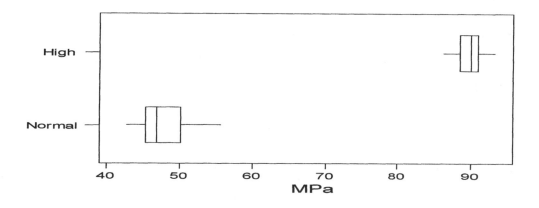

The most significant feature of these boxplots is the fact that their locations (medians) are far apart.

(b) This data is paired because the two measurements are taken for each of 15 test conditions. Therefore, we have to work with the differences of the two samples. A quantile plot of the 15 differences shows that the data follows (approximately) a straight line, indicating that it is reasonable to assume that the differences follow a normal distribution. Taking differences in the order 'Normal' - 'High', we find $\bar{d} = -42.23$ and $s_d = 4.34$. A 95% confidence interval for the difference between the population means is given by $\bar{d} \pm t \, {}^{s_d}\!/\!\sqrt{n}$, where n = 15 is the number of pairs and the critical t value is t = 2.145, based on n-1 = 14 df. The desired interval is: $-42.23 \pm (2.145) \, {}^{4.34}\!/\!\sqrt{15} = -42.23 \pm$ 2.404 = [-44.63, -39.83]. Because 0 is not contained in this interval, we can conclude that the difference between the population means is not 0; i.e., we conclude that the two population means are not equal.

55. This is paired data with n = 60, $\bar{d} = 4.47$, and $s_d = 8.77$. The critical t value associated with df = n-1 = 60-1 = 59 and 99% confidence is approximately 2.66 (Table IV). Alternatively, since df = 59 is large, we could simply use the 99% z value of 2.58, which we do in the following calculation: $\bar{d} \pm t \, {}^{s_d}\!/\!\sqrt{n} = 4.47 \pm (2.58) \, {}^{8.77}\!/\!\sqrt{60} = 4.47 \pm 2.92 = [1.55, 7.39]$.

Therefore, we estimate that the average blood pressure in a dental setting exceeds the average blood pressure in a medical setting by between 1.55 and 7.39. The interval does not contain 0, which suggests that the true average pressures are indeed different in the two settings.

57. (a) To generate a bootstrap interval using a Minitab macro, use the following types of Minitab commands in the macro (for data stored in column c1; set a counter k1 = 1 to start):

MTB> sample 20 c1 c10; ← (draws a random sample of size 20)

MTB> replace. ← (this assures sampling is done with replacement)

MTB> let c2(k1) = mean(c10) ← (store sample mean in column c2)

MTB> let k1 = k1 +1 ← (advance counter by 1)

Run your macro any number of times, say 200 or more and then use the stored results in column c2 to form the bootstrap interval. Our interval, based on 200 runs of the above statements gave a 95% bootstrap interval of [.8855, .9590]. Your results may be slightly different since different random samples will be used each time a macro is run. *Note: the answer in the text mistakenly gives a bootstrap interval for the data in Exercise 39 of Chapter 7.*

(b) The t confidence interval in Exercise 43 is [.888, .963] is very close to the bootstrap Interval [.886, .959] found in part (a).

59.   (a) As we showed in Example 7.15, the sample proportion $p = x/n$ is a maximum likelihood estimator of the population proportion π. That is, the MLE for π is just x/n.

(b) In Section 5.6, we showed that the mean of the sampling distribution of p coincides with π: i.e., $\mu_p = \pi$. Therefore, x/n (the sample proportion) is an unbiased estimator of π.

(c) As shown in part (b) above, x/n is the MLE of π. Using the function $g(\theta) = (1-\theta)^5$ and the Invariance Property of MLEs: since x/n is the MLE for π, then $g(x/n)$ is the MLE for $g(\pi)$. That is, $(1-x/n)^5$ is the MLE for $(1-\pi)^5$.

61.   (a) Denote the 95th percentile of a normal distribution by $x_{.95}$. By definition, then, the area to the left of $x_{.95}$ under the normal curve must be .95; i.e., $P(x \leq x_{.95}) = .95$. For a standard normal distribution, the z value with a cumulative area of .95 is approximately 1.645, which is 1.645 standard deviations (since $\sigma = 1$ for the z distribution) to the right of the mean (which is 0 for the z distribution). Therefore, for any normal distribution the value of $x_{.95}$ is 1.645 standard deviations to the right of its mean, so $x_{.95} = \mu + 1.645\sigma$. Next, we know from Example 7.17 that $\bar{x}$ and s^* are the MLEs for μ and σ for a normal distribution (*here* $s^* = s\sqrt{\frac{n-1}{n}}$, where s is the sample standard deviation of the data). Using the function $g(\theta_1, \theta_2) = \theta_1 + 1.645\theta_2$ and the Invariance Property of MLEs, we can then conclude that $g(\bar{x}, s^*) = \bar{x} + 1.645s^*$ is the MLE for $x_{.95}$.

(b) $x_{.95} \approx \bar{x} + 1.645s^* = 384.4 + (1.645)(19.879)\sqrt{10 - \frac{1}{10}} = 384.4 + 31.03 = 415.4.$

63.   (a) The likelihood function is $L(\lambda,\theta) = \lambda^n e^{-\lambda \sum_1^n (x_i - \theta)}$. After a little simplification, we find

$\frac{\partial L}{\partial \theta} = n\lambda^{n+1} e^{-\lambda \sum_1^n (x_i - \theta)}$. Note that this expression is always positive (for $\lambda > 0$), so there

is <u>no</u> value of θ that will satisfy $\frac{\partial L}{\partial \theta} = 0$. Instead, we have to take another approach to

maximizing $L(\lambda,\theta)$. Taking the partial derivative $\frac{\partial L}{\partial \lambda}$, after much simplification we

find:

$\frac{\partial L}{\partial \lambda} = \lambda^{n-1} e^{-\lambda \sum_1^n (x_i - \theta)} [n - \lambda \sum_i^n (x_i - \theta)]$, which, when set equal to 0, yields the solution

$\lambda = \frac{1}{\bar{x} - \theta}$.   (*Note: only the part in the brackets can equal 0, so we just solve*

$n - \lambda \sum_i^n (x_i - \theta) = 0$ *for* λ.) Rewriting $L(\lambda,\theta)$ as $\lambda^n e^{-\lambda n \bar{x}} e^{\lambda n \theta}$, we notice that θ only

appears in the expression $e^{\lambda n \theta}$, so the value of θ that maximizes $L(\lambda,\theta)$ will have to be

one that maximizes $e^{\lambda n \theta}$. Since <u>all</u> data points must exceed θ, then $\theta \le \min(x_1, x_2, ...,$

$x_n)$, so the largest possible value of θ that is allowed by the data is $\theta = \min(x_1, x_2, ..., x_n)$.

Notice that this value of θ does not depend on the particular value of λ, so we can

conclude that $\hat{\theta} = \min(x_1, x_2, ..., x_n)$ is the MLE of θ. Substituting in to the expression

for λ, the MLE of λ is $\hat{\lambda} = \frac{1}{\bar{x} - \hat{\theta}}$.

(b) $\hat{\theta} = \min(x_1, x_2, ..., x_n) = 0.64$ and $\hat{\lambda} = \frac{1}{\bar{x} - \hat{\theta}} = \frac{1}{5.58 - .64} = 0.202.$

65. The value $\lambda = 2$ is too large. It essentially puts normal curves with standard deviations of 2s around each data point. That is, normal curves with standard deviations that are twice the standard deviation of the data points themselves. The resulting kernel density will not show much detail in the data.

67. (a) The standard deviation used for the normal curves around each data point will equal $\lambda s = (d/3s)s = d/3$, where d is the minimum distance between any two of the data points. Therefore, the 3-sigma values for these normal curves will equal $\pm 3\sigma = \pm 3(d/3) = \pm d$, so the curves will be so closely packed around each data point that their 3σ values will usually not even overlap. This will lead to a density curve with a very choppy appearance; just a bunch of very small non-overlapping bell curves sitting over the data points.

(b) Larger values of λ will result in smoother kernel density curves.

69. λ will have to be raised. The reason is that the new sample standard deviation, s_{new} , will be slightly smaller than the original standard deviation s_{orig} since the data point removed was an outlier. That is, $s_{orig}/s_{new} > 1$. Therefore, to make the standard deviations used to create the kernel densities approximately equal, $\lambda_{orig}s_{orig} \approx \lambda_{new}s_{new}$, we must have $\lambda_{new}/\lambda_{orig} \approx s_{orig}/s_{new} > 1$.

71. (a) For these n =153 observations, $\bar{x} = 135.39$ and s = 4.59. For a 95% lower confidence bound, $\alpha = .05$ (not .025, since this is a one-sided interval). The interval is given by:

$$\bar{x} - z_{\alpha}\frac{s}{\sqrt{n}} < \mu$$

$$135.39 - 1.645\frac{4.59}{\sqrt{153}} < \mu$$

$$134.78 < \mu$$

(b) In order to use the interval in (a) it is necessary to assume that the population of tensile strengths is approximately normal, although the Central Limit theorem also validates the use of interval when the sample size is large (which is the case here).

(c) A Minitab histogram of this data appears approximately bell-shaped, so the normality assumption in part (b) is a good one for this data.

(d)   A 95% lower <u>prediction</u> bound on the tensile strength of the next item selected is:

$$\bar{x} \; - \; t_\alpha s \sqrt{1 + \frac{1}{n}} \; < \; \mu$$

$$135.39 \; - \; 1.645(4.59)\sqrt{1 + \frac{1}{153}} \; < \; \mu$$

$$127.81 \; < \; \mu$$

73.   The center of any confidence interval for μ_1-μ_2 is always $\bar{x}_1$- $\bar{x}_2$, so $\bar{x}_1$- $\bar{x}_2$ = (-473.3 + 1691.9)/2 = 609.3. Furthermore, the half-width of this interval is [1691.9-(-473.3)]/2 = 1082.6. Equating this value to the expression for the half-width of a 95% interval, 1082.6 = $(1.96)\sqrt{\frac{s_1^2}{n_1} + \frac{s_2^2}{n_2}}$, we find $\sqrt{\frac{s_1^2}{n_1} + \frac{s_2^2}{n_2}}$ = 1082.6/1.96 = 552.35. For a 90% interval, the associated z value is 1.645, so the 90% confidence interval is then

$$\bar{x}_1\text{- } \bar{x}_2 \pm (1.645)\sqrt{\frac{s_1^2}{n_1} + \frac{s_2^2}{n_2}} = 609.3 \pm (1.645)(552.35) = 609.3 \pm 908.6 = [-299.3, 1517.9].$$

75. $n_1 = n_2 = 40$, $\bar{x}_1 = 3975.0$, $s_1 = 245.1$, $\bar{x}_2 = 2795.0$, $s_2 = 293.7$. The large-sample 99%

confidence interval for μ_1-μ_2 is then: $\bar{x}_1$- $\bar{x}_2 \pm (1.645)\sqrt{\frac{s_1^2}{n_1} + \frac{s_2^2}{n_2}} =$

$$(3975.0 - 2795.0) \pm (2.58)\sqrt{\frac{245.1^2}{40} + \frac{293.7^2}{40}} = 1180.0 \pm 156.05 \approx [1024, 1336]. \text{ The value}$$

0 is not contained in this interval so we can state that, with very high confidence, the value of μ_1-μ_2 is not 0, which is equivalent to concluding that the population means are not equal.

77.   (a) The sample proportion is p = 38/250 = .152. The z-value for a 95% confidence interval

for π is z = 1.96, so the desired confidence interval is $p \pm z \sqrt{\frac{p(1-p)}{n}} =$

$$.152 \pm (1.96)\sqrt{\frac{.152(1-.152)}{250}} = .152 \pm .045 = [.107, .197].$$

(b) The 'new' value for p is $(38+2)/(250+2) = .1587$ and the revised confidence interval is

$$p \pm z \sqrt{\frac{p(1-p)}{n}} = .1587 \pm (1.96) \sqrt{\frac{.1587(1-.1587)}{252}} = .1587 \pm .0451 = [.114, .204].$$

The intervals in (a) and (b) differ slightly.

79. (a) By the definition of the median, $P(x_1 < \tilde{\mu}) = .5$. Of course, $P(x_1 > \tilde{\mu}) = .5$ too, a fact that we use in part (d) below.

(b) P(both observations are smaller than median) $= P(x_1 < \tilde{\mu}$ and $x_2 < \tilde{\mu})$

$= P(x_1 < \tilde{\mu}) \cdot P(x_2 < \tilde{\mu}) = (.5)(.5) = .25$. Note that we have used the fact that random sampling guarantees that the random variables x_1 and x_2 are independent, which reduces the probability calculation to a simple multiplication.

(c) $y_n = \max(x_1, x_2, x_3, \ldots, x_n)$, so $P(y_n < \tilde{\mu}) = P(\underline{\text{all}} \ x_i\text{'s are less than } \tilde{\mu})$

$= P(x_1 < \tilde{\mu}) \cdot P(x_2 < \tilde{\mu}) \cdots P(x_n < \tilde{\mu}) = (.5)^n$. As in part (b), independence allows us to multiply the separate probabilities.

(d) $y_1 = \min(x_1, x_2, x_3, \ldots, x_n)$, so $P(y_1 > \tilde{\mu}) = P(\text{all } x_i\text{'s are greater then } \tilde{\mu})$

$= P(x_1 > \tilde{\mu}) \cdot P(x_2 > \tilde{\mu}) \cdots P(x_n > \tilde{\mu}) = (.5)^n$.

(e) $P(y_1 < \tilde{\mu} < y_n) = 1 - P(\text{either } y_n < \tilde{\mu} \text{ or } y_1 > \tilde{\mu}) = 1 - [P(y_n < \tilde{\mu}) + P(y_1 > \tilde{\mu})]$

$= 1 - [(.5)^n + (.5)^n] = 1 - (.5)^{n-1}$. Note that the events $y_n < \tilde{\mu}$ and $y_1 > \tilde{\mu}$ are disjoint, so we <u>can</u> simply use the addition law. Regarding $[y_1, y_n]$ as a confidence interval for $\tilde{\mu}$, we can say that this interval has an associated confidence level of $1 - (.5)^{n-1}$.

(f) For this data, $y_1 = 28.7$ and $y_n = 42.0$, so $[28.7, 42.0]$ is a confidence interval for $\tilde{\mu}$ that has $1-(.5)^{10-1} = .998$, or 99.8 confidence.

(g) The sample mean and standard deviation of the data are 34.45 and 4.2914, respectively. Based on n-1 = 9 df, the critical t value for a two-sided 99.8% confidence interval is 4.297 (from Table IV), so the desired interval is $\bar{x} \pm t \cdot s/\sqrt{n} = 34.45 \pm (4.297) \cdot 4.2914/\sqrt{10}$

34.45 ± 5.83 = [28.62, 40.28]. This interval is a little narrower than the one in (f), which is usually the case as long as the assumption that we are sampling from a normal distribution is valid. Note that the interval in (e) and (f) is valid <u>regardless</u> of the distribution of the population values.

81. (a) x_{n+1} is as likely to be above x_1 as below it, so $P(x_{n+1} > x_1) = .50$.

(b) Any of the three values has an equal chance of being the smallest, so $P(x_{n+1}$ is smallest$) = 1/3$.

(c) Using the same reasoning as in (b), $P(x_{n+1} < y_1) = P(x_{n+1}$ is the smallest of n random observations$) = 1/(n+1)$. Likewise, $P(x_{n+1} > y_n) = P(x_{n+1}$ is the largest of n random observations$) = 1/(n+1)$.

(d) $P(y_1 < x_{n+1} < y_n) = P(x_{n+1}$ is neither the largest nor smallest value$) = 1 - P(x_{n+1}$ is the smallest or x_{n+1} is the largest$) = 1 - [1/(n+1) + 1/(n+1)] = 1 - 2/(n+1)$. Therefore, the interval $[y_1, y_n]$ is a <u>prediction</u> interval (since it estimates where a single data value is, not where the mean is) with an associated confidence level of $1 - 2/(n+1)$. For the data of Exercise 77, the interval [28.7, 42.0] would be a $1 - 2/(n+1) = 1 - 2/(10+1) = .818$, or 81.8% prediction interval for the <u>next</u> curing time.

83. The following normal quantile plot shows three points that fall far below a relatively straight line formed by the remaining points. This pattern indicates that population normality is very implausible. *Note :to review the construction of normal quantile plots, refer to the answer to Exercise 45of Chapter 2.*

Percent oil

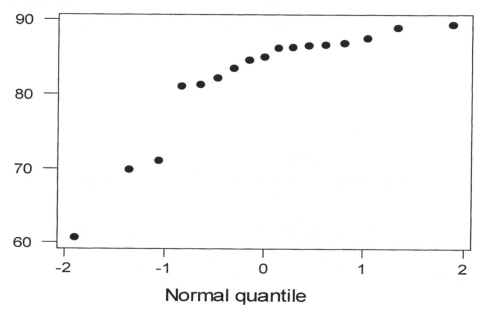

In Exercise 80, a confidence interval for the median was based on the second largest and second smallest observations was shown to have a confidence level of $1-2(n+1)(.5)^n$. For the data of Exercise 83, this yields the interval [69.80, 88.80], which has an associated confidence level of $1 - 2(17+1)(.5)^{17} = .9997$, or 99.97%.

85.   The desired point estimate is $\hat{\theta} = \bar{x}_1 - (\bar{x}_2 + \bar{x}_3)/2 = 38,376 - (41,569+42,123)/2 = -3470$.

In this example, $a_1 = 1$, $a_2 = -\frac{1}{2}$, $a_3 = -\frac{1}{2}$, so $s_{\hat{\theta}}^2 = (1)^2\dfrac{(1522)^2}{40} + (-\frac{1}{2})^2\dfrac{(1711)^2}{32} +$

$(-\frac{1}{2})^2\dfrac{(1645)^2}{32} = 101,924.18$, so $s_{\hat{\theta}} = 319.26$. A 95% confidence interval is then $-3470 \pm$

$(1.96)(319.26) = -3470 \pm 625.75 \approx [-4096, -2844]$. Therefore, we can be 95% confident that the average of the two national brands exceeds the store brand average by between 2844 and 4096 miles.

87.   (a) Letting z_p denote the value from the standard normal distribution for which p% of the area lies to left of z_p, the pth percentile of a normal distribution can be written as $\mu + z_p\sigma$. Putting this expression in for "z percentile" in the noncentral t formula and using the fact that, for 95% confidence, $t_{.025} < t < t_{.975}$, we can rearrange the inequality to obtain the following confidence interval for the pth percentile:

$$\bar{x} - t_{.975} \frac{s}{\sqrt{n}} \; < \; \mu + z_p\sigma \; < \; \bar{x} - t_{.025} \frac{s}{\sqrt{n}}$$

(b) Using the values given in part (b) of the problem, a 95% confidence interval for the 5th percentile of the modulus of elasticity distribution is given by:

$14532.5 - (10.9684)(2055.67)/\sqrt{16} \; < 5^{th}$ percentile $< 14532.5 - (4.1690)(2055.67)/\sqrt{16}$
which simplifies to [8895.65, 12,389.98].

Chapter 8
Testing Statistical Hypotheses

1. (a) Yes, $\sigma > 100$ is a statement about a population standard deviation, i.e., a statement about a population parameter.

  (b) No, this is a statement about the statistic $\bar{x}$, not a statement about a population parameter.

  (c) Yes, this is a statement about the population median $\tilde{\mu}$.

  (d) No, this is a statement about the statistic s (s is the <u>sample</u>, not population, standard deviation.

  (e) Yes, the parameter here is the ratio of two other parameters; i.e, σ_1/σ_2 describes some aspect of the populations being sampled, so it is a parameter, not a statistic.

  (f) No, saying that the difference between two samples means is -5.0 is a statement about sample results, not about population parameters.

  (g) Yes, this is a statement about the parameter λ of an exponential population.

  (h) Yes, this is a statement about the proportion π of successes in a population.

  (i) Yes, this is a statement about the form of a population distribution.

  (j) Yes, this is a statement about the parameter values of a specific distribution.

3.   Let μ denote the average amperage in the population of all such fuses. Then the two relevant hypotheses are $H_0: \mu = 40$ (the fuses conform to specifications) and $H_a: \mu \neq 40$ (the average amperage either exceeds 40 is less than 40).

5.   Let μ denote the average breaking distance for the new system. The relevant hypotheses are $H_0: \mu = 120$ versus $H_a: \mu < 120$, so implicitly H_0 really says that $\mu \geq 120$. A Type I error would be: *concluding that the new system really does reduce the average breaking distance (i.e., rejecting H_0) when, in fact (i.e., when H_0 is true) it doesn't.* A Type II error would be: *concluding that the new system does not achieve a reduction in average breaking distance (i.e., not rejecting H_0) when, in fact (i.e, when H_0 is false) it actually does.*

7.   Let μ_1 denote the true average warpage for the regular laminate and let μ_2 denote the true average warpage for the special laminate. The hypotheses of interest are $H_0: \mu_1 - \mu_2 = 0$ versus $H_a: \mu_1 - \mu_2 > 0$, so implicitly H_0 asserts that $\mu_1 - \mu_2 \leq 0$, that is, that the regular laminate does at least as well or better than the special laminate. A Type I error would be: *concluding that the special laminate outperforms the regular laminate (i.e., rejecting H_0) when, in fact (i.e., when H_0 is true) this is not the case.* A Type II error is: *concluding that the special laminate is not better than the regular laminate (i.e., not rejecting H_0) when, in fact (i.e., when H_0 is false) the special laminate really does outperform the regular laminate.*

9.   (a) $.001 = $ P-value $\leq \alpha = .05$, so H_0 should be rejected.

        (b) $.021 = $ P-value $\leq \alpha = .05$, so H_0 should be rejected.

        (c) $.078 = $ P-value $> \alpha = .05$, so H_0 should <u>not</u> be rejected.

        (d) $.047 = $ P-value $\leq \alpha = .05$, so H_0 should be rejected.

        (e) $.156 = $ P-value $> \alpha = .05$, so H_0 should <u>not</u> be rejected.

11. (a) $z = \dfrac{\bar{x}-5}{s/\sqrt{n}} = \dfrac{5.23-5}{.89/\sqrt{50}} = 1.83$, so P-value = area under the z curve to the right (because

it's a right-tailed test) of z = 1.83. Therefore, P-value = P(z>1.83) = P(z<-1.83) = .0336.

(b) $z = \dfrac{\bar{x}-5}{s/\sqrt{n}} = \dfrac{5.72-5}{1.01/\sqrt{35}} = 4.22$, so P-value = area under the z curve to the right of z = 4.22.

Therefore, P-value = P(z>4.22) ≈ 0.

(c) $z = \dfrac{\bar{x}-5}{s/\sqrt{n}} = \dfrac{5.35-5}{1.67/\sqrt{40}} = 1.33$, so P-value = area under the z curve to the right of z = 1.33.

Therefore, P-value = P(z>1.33) = P(z<-1.33) = .0918.

13. (a) This is a test about the population average μ = average silicon content in iron. The null

hypothesis value of interest is μ = .85 so the test statistic is of the form $z = \dfrac{\bar{x}-.85}{s/\sqrt{n}}$.

From the wording of the exercise it seems that a 2-sided test is appropriate (since the
silicon content is supposed to average .85 and not be substantially larger or smaller than
that number), so the relevant hypotheses are H_0:μ = .85 versus H_a:μ ≠ .85 . We can
verify that 2-sided test was done by calculating the P-value associated with the z value of
-.81 given in the printout: the area to the left of -.81 is .2090 (from Table I), so the
2-sided P-value associated with z = -.81 is 2(.2090) = .418 ≈ .42.

(b) The P-value of .42 is quite large, so we don't expect it to lead to rejecting H_0 for any of
the usual values of α used in hypothesis testing. Indeed, P = .42 exceeds both α = .05
and α = .10, so in neither case would this data lead to rejecting H_0. It appears to be quite
likely that the average silicon content does not differ from .85

15. Let μ denote the true average speedometer reading (at 55 mph). Since we are concerned
about whether the speedometer readings may be too high or too low (compared to 55 mph),
this requires a 2-sided test of H_0:μ = 55 versus H_a: μ ≠ 55. The test statistic would be z =

$\dfrac{\bar{x}-\mu_0}{s/\sqrt{n}} = \dfrac{53.87-55}{1.36/\sqrt{40}} = -5.25$. Therefore, P-value = 2P(z < -5.25) ≈ 2(.0000) = 0. Since this

P-value is certainly less than $\alpha = .01$, we reject H_0 and conclude that the average reading is not equal to 55. In particular, the average reading is below 55 so there is a problem with the calibration of the speedometers.

17. (a) P-value = area under the t curve (with d.f. = 8) to the right of t = 2.0. From Table VI, this tail area is .040.

(b) P-value = area under the t curve (with d.f. = 11 to the left of t = -2.4. From Table VI, this tail area is .018. Note that we have used the fact that $P(t < -2.4) = P(t > +2.4)$ in order to use Table VI.

(c) P-value = twice the area under the t curve (with d.f. = 15) to the left of -1.6. From Table VI, this area is $2(.065) = .13$.

(d) P-value = area under the t curve (with d.f. = 19) to the right of t = -.4. That is,
$P(t > -.4) = P(-.4 < t < 0) + P(t \geq 0) = [.5 - P(t < -.4)] + P(t \geq 0) = [.5 - .347] + .5 = .653$.

(e) P-value = area under the t curve (with d.f. = 5) to the right of t = 5.0. Note that t = 5.0 is not in Table VI, which means that the tail area $P(t > 5.0)$ is approximately 0.

(f) P-value = twice the area under the t curve (with d.f. = 40) to the left of -4.8. As in part (e), t = 4.8 is not contained in Table VI, so P-value = $2P(t < -4.8) = 2P(t > 4.8) \approx 2(.0000)$ = 0.

19. (a) Let μ denote the true average writing lifetime. The wording in this exercise indicates that the investigators believe, a priori, that μ can't be less than 10 (i.e., $\mu \geq 10$), so the relevant hypotheses are H_0: $\mu = 10$ versus H_a: $\mu < 10$.

(b) The degrees of freedom are d.f. = n-1 = 18-1 = 17, so the P-value = $P(t < -2.3) = P(t > 2.3) = .017$. Since P-value = $.017 < \alpha = .05$, we should reject H_0 and conclude that the design specification has not been satisfied.

(c) For t = -1.8, P-value = P(t<-1.8) = P(t>1.8) = .045, which exceeds α = .01. Therefore, in This case H_0 would not be rejected. That is, there is not sufficient evidence to conclude that the design specification is not satisfied.

(d) For t = -3.6, P-value = P(t<-3.6) = P(t>3.6) = .001. This P-value is smaller than either α = .01 or α = .05, so in either case H_0 would be rejected. In fact, H_0 would be rejected for any value of α that exceeds .001. For such values of α, we would conclude that the design specification has not been satisfied.

21. (a) The lower quartile (i.e., the left-hand edge of the boxplot) lies above 200, so we can see that more than 3/4 or 75% of the sample observations lie above 200. Thus, it appears very doubtful that the design specification has been satisfied.

(b) Let μ denote the true average coating weight. The appropriate hypothesis test is 2-sided (since coating weights that are too large or too small are undesirable), so we test H_0:μ = 200 versus H_a:$\mu \neq$ 200. The test statistic is $t = \dfrac{\bar{x} - \mu_0}{s/\sqrt{n}} = \dfrac{206.73 - 200}{6.35/\sqrt{30}} = 5.80$ with d.f. = n-1 = 30 -1 = 29. From Table VI, the 2-sided P-value is 2P(t > 5.80) $\approx$ 2(.0000) = 0. Therefore, H_0 should be rejected, confirming the conclusion drawn in part (a).

23. From the wording in this exercise the value 3000 is a target value for viscosity, so we don't want to have the true average viscosity μ be smaller or larger than 3000, which means that a 2-sided test is needed. To test H_0:μ = 3000 versus H_a:$\mu \neq$3000, the test statistic is $t = \dfrac{\bar{x} - \mu_0}{s/\sqrt{n}}$

$= \dfrac{2887.6 - 3000}{84.026/\sqrt{5}} = -2.99 \approx - 3.0$. Using d.f. =n-1 = 5-1 = 4, Table VI gives the 2-sided P-value of 2P(t < -3.0) = 2P(t > 3.0) = 2(.020) = .040. For any value of α greater than .040 (such as α = .05) we would reject H_0, but we would not reject H_0 for values of α smaller than .040.

25. (a) df $\approx \dfrac{\left[\frac{5.0^2}{10} + \frac{6.0^2}{10}\right]^2}{\left[\frac{5.0^2}{10}\right]^2\big/9 + \left[\frac{6.0^2}{10}\right]^2\big/9} = 17.43$, so round <u>down</u> to df ≈ 17.

(b) df $\approx \dfrac{\left[\frac{5.0^2}{10} + \frac{6.0^2}{15}\right]^2}{\left[\frac{5.0^2}{10}\right]^2\big/9 + \left[\frac{6.0^2}{15}\right]^2\big/14} = 21.71$, so round <u>down</u> to df ≈ 21.

(c) df $\approx \dfrac{\left[\frac{2.0^2}{10} + \frac{6.0^2}{15}\right]^2}{\left[\frac{2.0^2}{10}\right]^2\big/9 + \left[\frac{6.0^2}{15}\right]^2\big/14} = 18.27$, so round <u>down</u> to df ≈ 18.

(d) df $\approx \dfrac{\left[\frac{5.0^2}{12} + \frac{6.0^2}{24}\right]^2}{\left[\frac{5.0^2}{12}\right]^2\big/11 + \left[\frac{6.0^2}{24}\right]^2\big/23} = 26.08$, so round <u>down</u> to df ≈ 28.

27. Let μ_1 denote the true average strength for wire-brushing preparation and let μ_2 denote the average strength for hand-chisel preparation. Since we are concerned about any possible difference between the two means, a 2-sided test is appropriate, so we test $H_0: \mu_1-\mu_2 = 0$ versus $H_a: \mu_1-\mu_2 \neq 0$ using the results given in the exercise: $n_1 = 12$, $\bar{x}_1 = 19.20$, $s_1 = 1.58$, $n_2 = 12$, $\bar{x}_2 = 23.13$, $s_2 = 4.01$. The test statistic is:

$$t = \frac{\bar{x}_1 - \bar{x}_2}{\sqrt{\frac{s_1^2}{n_1} + \frac{s_1^2}{n_2}}} = \frac{19.20 - 23.13}{\sqrt{\frac{1.58^2}{12} + \frac{4.01^2}{12}}} = -3.93/1.244 = -3.16 \approx -3.2.$$

The approximate d.f. $\approx \dfrac{\left[\frac{1.58^2}{12} + \frac{4.01^2}{12}\right]^2}{\left[\frac{1.58^2}{12}\right]^2\big/11 + \left[\frac{4.01^2}{12}\right]^2\big/11} = 2.396/.1672 = 14.34$, which we round

<u>down</u> to d.f. = 14. For a 2-sided test, we use Table VI to find the P-value = $2P(t < -3.2) = 2(.006) = .006$. Since P-value = .006 is less than $\alpha = .05$, we reject H_0 and conclude that there does appear to be a difference between the two population average strengths.

29. Let μ_1 denote the true average proportional stress limit for red oak and let μ_2 denote the average stress limit for Douglas fir. The wording of the exercise suggests that we are interested in detecting any differences between the two averages, which means a 2-sided test is appropriate, so we test H_0: $\mu_1-\mu_2 = 0$ versus H_a: $\mu_1-\mu_2 \neq 0$. The test statistic is:

$$t = \frac{\bar{x}_1 - \bar{x}_2}{\sqrt{\frac{s_1^2}{n_1} + \frac{s_1^2}{n_2}}} = \frac{8.48 - 6.65}{\sqrt{\frac{.79^2}{14} + \frac{1.28^2}{10}}} = 1.83/.4565 = 4.01 \approx 4.0.$$

The approximate d.f. $\approx \dfrac{\left[\frac{.79^2}{14} + \frac{1.28^2}{10}\right]^2}{\dfrac{\left[\frac{.79^2}{14}\right]^2}{13} + \dfrac{\left[\frac{1.28^2}{10}\right]^2}{9}} = .04344/.003135 = 13.85$, which we round

down to d.f. = 13. For a 2-sided test, we use Table VI to find the P-value = $2P(t > 4.0) = 2(.001) = .002$. Since P-value = .002 is very small (smaller than the usual significance like .05 or .01), we reject H_0 and conclude that there is a difference between the two average stress limits.

31. (a) A point estimate for $\mu_{treatment}$ - $\mu_{control}$ is given by $\bar{x}_{treatment}$ - $\bar{x}_{control}$ = $19.9 - 13.7 = 6.2$ days. This suggests that there might be a difference of about 6.2 days between the two groups, but there is no way to be sure if this result is significant until you do a hypothesis test (which takes compares this estimate to the magnitude of the sample-to-sample variation expected in such an estimate).

(b) Since the ratio of the largest to smallest sample variance exceeds 4, it is best to use the unpooled t-test. For a two-sided test of H_0:$\mu_t-\mu_c = 0$ versus H_a: $\mu_t-\mu_c \neq 0$, we calculate the test statistic:

$$t = \frac{\bar{x}_t - \bar{x}_c}{\sqrt{\frac{s_t^2}{n_t} + \frac{s_c^2}{n_c}}} = \frac{19.9 - 13.7}{\sqrt{\frac{39.1^2}{60} + \frac{15.8^2}{60}}} \approx 1.14.$$

The standard errors of the two statistics are $se_t = 39.1/\sqrt{60} = 5.0478$ and $se_c = 15.8/\sqrt{60} = 2.0398$

The approximate d.f. $= \dfrac{\left[(se_1)^2 + (se_2)^2\right]^2}{\dfrac{(se_1)^4}{n_1 - 1} + \dfrac{(se_2)^4}{n_2 - 1}} = \dfrac{\left[(5.0478)^2 + (2.0398)^2\right]^2}{\dfrac{(5.0478)^4}{60 - 1} + \dfrac{(2.0398)^4}{60 - 1}} = 77.8,$

which we round <u>down</u> to 77. Using, say, $\alpha = .05$, the critical values are $t_{.025}(77\text{df}) \approx$ 1.991 (using Minitab or Excel). Since $t = 1.14$ doesn't exceed 1.991, this data does <u>not</u> show that there is a significant difference between the means of the treatment and the control groups.

(c) No, the ICU stay for the treatment group does <u>not</u> appear to follow a normal distribution because the mean 19.9 is too close to zero, as measured in standard deviations: $z = (0-19.9)/39.1 = .509$. Since ICU stay times must be nonnegative, the normal distribution would require that a large percentage of the stay times be negative, which is impossible.

33. Let μ_1 denote the true average weight gain for steroid treatment and let μ_2 denote the average weight gain for the population not treated with steroids. The exercise asks if we can conclude that μ_2 exceeds μ_1 by more than 5 g., which we can restate in the equivalent form: $\mu_1 - \mu_2 < -5$. Therefore, we conduct a 1-sided test of $H_0: \mu_1 - \mu_2 = -5$ versus $H_a: \mu_1 - \mu_2 < -5$. The test statistic is:

$$t = \dfrac{\bar{x}_1 - \bar{x}_2}{\sqrt{\dfrac{s_1^2}{n_1} + \dfrac{s_1^2}{n_2}}} = \dfrac{32.8 - 40.5 - (-5)}{\sqrt{\dfrac{2.6^2}{8} + \dfrac{2.5^2}{10}}} = -2.7/1.2124 = -2.23 \approx -2.2.$$

The approximate d.f. $\approx \dfrac{\left[\dfrac{2.6^2}{8} + \dfrac{2.5^2}{10}\right]^2}{\dfrac{\left[\dfrac{2.6^2}{8}\right]^2}{7} + \dfrac{\left[\dfrac{2.5^2}{10}\right]^2}{9}} = 2.1609/.1454 = 14.86,$ which we round

<u>down</u> to d.f. $= 14$. For a lower-tailed test, we use Table VI to find the P-value $= P(t<-2.2) = P(t>2.2) = .022$. Since this P-value is larger than the specified significance level $\alpha = .01$, we can not reject H_0. Therefore, this data does not support the belief that average weight gain for the control group exceeds that of the steroid group by more that 5 g.

35. Each pair of brothers are tested and the data refers to the difference between the two measurements in each pair of brothers. Let μ_d denote the average difference between the readings for DES exposure versus no exposure to DES. Since we are concerned with detecting a possible drop in spatial ability, the appropriate hypotheses are $H_0: \mu_d = 0$ versus

$H_a: \mu_d < 0$ (note: we assume that the each data point equal the difference of the DES exposed brother minus that for the non exposed brother). The test statistic is:

$$t = \frac{\bar{d} - 0}{s_d / \sqrt{n}} = \frac{\bar{x}_1 - \bar{x}_2 - 0}{s_d / \sqrt{n}} = \frac{12.6 - 13.8 - 0}{.5} = \text{-2.4.} \quad \text{For d.f.} = n\text{-}1 = 10\text{-}1 = 9, \text{ the P-value for}$$

this lower-tailed test is P(t<-2.4) = P(t>2.4) = .020, which is smaller than the specified significance level of $\alpha = .05$, so H_0 is rejected. Therefore, if $\alpha = .05$ we can conclude that DES exposure does result in a reduced spatial ability. However, since P-value = .02 exceeds .01, we could not make the same statement if α were changed from .05 to .01.

37.   Each sample is measured by two different methods so the data is paired. The observed differences (MSI minus SIB) are: 0.03, -0.51, -0.80, -0.57, -0.66, -0.63, -0.18, 0.01. The sample mean and standard deviation of these observations are $\bar{d} = -.4138$ and $s_d = .3210$. The wording in the exercise indicates that we are interested in detecting *any* difference between the two methods, so a 2-sided test is required. To test $H_0: \mu_d = 0$ versus $H_a: \mu_d \neq 0$, the test statistic is:

$$t = \frac{\bar{d} - 0}{s_d / \sqrt{n}} = \frac{-.4138 - 0}{.3210 / \sqrt{8}} = \text{-3.64} \approx \text{-3.6.}$$

For d.f. = n-1 = 8-1 = 7, we use Table VI to find the P-value for this 2-sided test: P-value = 2P(t < -3.6) = 2(.004) = .008. At significance levels of either $\alpha = .05$ or $\alpha = .01$, H_0 would be rejected and we would conclude that there is a difference between the true average differences for the two methods. However, at $\alpha = .001$ we would not reject H_0 since the P-value of .008 exceeds .001.

39.   Refer to problem 34 (which references problem 51 of Chapter 7) for the discussion of the pooled t-test.

	T stat	df	p-value (2-sided)
Un-pooled t test:	-1.8072	15	.0908
Pooled t test:	-1.8926	24	.0705

41. (a) For d.f. $= 2$, the area under the chi-squared curve to the right of 7.82 is .025 and the area to the right of 7.37 is .030. Therefore, the area to the right of 7.5 lies somewhere between .025 and .030. That is, $.025 <$ P-value $< .030$.

(b) For d.f. $= 6$, the area under the chi-squared curve to the right of 13.19 is .04 and the area to the right of 12.87 is .0450. Therefore, the area to the right of 13.0 lies somewhere between .040 and .045. That is, $.040 <$ P-value $< .045$.

(c) For d.f. $= 9$, the area under the chi-squared curve to the right of 17.6 is .040 and the area to the right of 18.01 is .035. Therefore, the area to the right of 18.0 lies somewhere between .035 and .040. That is, $.035 <$ P-value $< .040$, although because 18.01 is so close to 18.0, we could also say that the P-value $\approx .035$.

(d) For d.f. $= 4$, the value 21.3 is not in the table, which means that the area under the chi-squared curve to the right of 21.3 is less than .001. That is, P-value $< .001$.

(e) For d.f. $= 3$, the value 5.0 is less than the first entry in the table..030, which means that the area to the right of 5.0 exceeds .10. That is, P-value $> .10$.

43. Using the numbering 1 (for Winter), 2 (Spring), 3(Summer), and 4 (Fall), let $\pi_i =$ the true proportion of all homicides committed in the ith season. If the homicide rate doesn't depend upon the season, then we would expect the rates to be equal; i.e., $H_0: \pi_1 = \pi_2 = \pi_3 = \pi_4 = .25$. The alternative hypothesis in this case would be H_a: *At least one of the proportions does not equal .25*. Using the proportions in H_0, the expected numbers of homicides (out of $n = 1361$) are shown below the actual numbers from the problem:

Season	Winter	Spring	Summer	Fall	total #
Observed	328	334	372	327	1361
Expected	340.25	340.25	340.25	340.25	1361

χ^2 contribution:	.4410	.1148	2.9627	.5160

The χ^2 test statistic value is $\chi^2 = .4410 + .1148 + 2.9627 + .5160 = 4.0345$. For d.f $= k-1 = 4-1$ $= 3$, the value 4.0345 is smaller than any of the entries in Table VII (column df=3), so the P-value associated with 4.0345 must be larger than .10. Therefore, since P-value $> .10 > .05 = \alpha$, H_0 is not rejected and we conclude that this data does not support the belief that there are different homicide rates in the different seasons.

45.   This is a χ^2 test of the homogeneity of several proportions. The hypotheses to test are H_0: *germination rate is homogenous across the different seed types* and H_a: *germination rate depends on the seed type*. The data (with expected values underneath) are shown below:

	1	2	Seed Type 3	4	5	Total
Germinated	31	57	87	52	10	237
	22.52	53.33	87.10	56.88	17.18	
Failed to germinate	7	33	60	44	19	163
	15.49	36.67	59.90	39.12	11.82	
Total	38	90	147	96	29	400

The χ^2 test statistic is $\chi^2 = $ $3.198 + 0.253 + 0.000 + 0.419 + 3.002$
$+4.649 + 0.368 + 0.000 + 0.609 + 4.365 = 16.864$.

For d.f. $= (5-1)(2-1) = 4$, the value of $\chi^2 = 16.864$ falls between entries 14.86 and 18.46 of Table VII, so we can say that $.001 <$ P-value $< .005$. Thus, the P-value is smaller that the specified significance level of $\alpha = .01$, so H_0 is rejected and we conclude that germination rates do depend upon seed type. More specifically, note that the largest χ^2 contributions are associated with seed types 1 and 5 (which account for most of the χ^2 value), so we conclude that seed type 1 shows a higher germination rate ($31/38 \approx 81.6\%$) than the rate for seed type 5 ($10/29 \approx 34/5\%$).

47.   (a) A sample from a single population has been selected so this is a test of independence of two factors, 'type of vehicle' and 'commuting distance'. That is, the relevant hypotheses are H_0:*vehicle type and commuting distance are independent* and H_a: *vehicle type and commuting distance are not independent.*

(b) The data (with expected values underneath them) for this χ^2 test is shown below:

| | **Commuting distance** | | | |
	0-<10	10-<20	≥ 20	Totals
Subcompact	6	27	19	52
	10.19	26.21	15.60	
Compact	8	36	17	61
	11.96	30.74	18.30	
Midsize	21	45	33	99
	19.40	49.90	29.70	
Full-size	14	18	6	38
	7.45	19.15	11.40	
Totals	49	126	75	250 = n

The χ^2 test statistic is χ^2 = Chi-Sq =   1.724 + 0.024 + 0.741 +
                                                  1.309 + 0.899 + 0.092 +
                                                  0.131 + 0.480 + 0.367 +
                                                  5.764 + 0.069 + 2.558 = 14.158

For d.f. = (3-1)(4-1) = 6, the value of χ^2 = 14.158 falls between entries 13.96 and 14.44 of Table VII, so we can say that .025 < P-value < .030. Thus, the P-value is smaller that the specified significance level of α = .05, so H_0 is rejected and we conclude that commuting distance and type of vehicle are related. Note that, of the contributions to the calculated χ^2 value, the largest ones occur in the 'commute less than 10 miles' column. Interpreting the data in that column alone, it appears that more full-size car owners than expected commute less than 10 miles, while fewer subcompact owners than expected commute less than 10 miles. A case can be made for the conclusion that owners of larger cars commute smaller distances than owners of small cars.

49. The pattern of points in the plot appear to deviate from a straight line, a conclusion that is also supported by the small P-value ($<$ .01000) of the Ryan-Joiner test. Therefore, it is implausible that this data came from a normal population. In particular, the observation 116.7 is a clear outlier. It would be dangerous to use the one-sample-t interval as a basis for inference.

51. The pattern appears to be approximately linear and the P-value for the Ryan-Joiner test is larger than .10, so we conclude that this data could reasonably have come from a normal population. This means that it would be legitimate to use a one-sample-t interval for inference with this data.

53. The equation $\mu = (1/\lambda) - \dfrac{x_0 e^{-\lambda x_0}}{1 - e^{-\lambda x_0}}$ can not be explicitly solved for λ. Instead, we replace μ and x_0 by the given values $\bar{x} = 13.086$ and $x_0 = 70$, then solve numerically for the estimated value $\hat{\lambda}$. That is, we solve the equation $13.086 = (1/\hat{\lambda}) - \dfrac{70 e^{-70\hat{\lambda}}}{1 - e^{-70\hat{\lambda}}}$ numerically for $\hat{\lambda}$.

The solution is $\hat{\lambda} \approx .0742$. With ($a_{i-1}$, a_i) denoting the ith class interval ($i = 1,2,3,\ldots,9$), the expected class frequencies are given by : (n)(proportion in i^{th} class) $= (40) \displaystyle\int_{a_{i-1}}^{a_i} f(x)dx =$

$40 \left[\dfrac{e^{-.0742 a_{i-1}} - e^{-.0742 a_i}}{1 - e^{-.0742(70)}} \right]$. The expected frequencies are: 18.0, 9.9, 5.5, 3.0, 1.8, .9, .5, .3, and .1. From these expected values we obtain the calculated χ^2 value of 1.34. Using d.f. $= k$ -1-m $= 9 - 1 - 1 = 7$, (m =1 because we estimated the single parameter λ from the data), the value 1.34 is much smaller than any of the entries in the df $= 7$ column of Table VII, so the P-value must exceed .10. In fact, the exact P-value is much larger still, so H_0 is not rejected and we conclude that it is reasonable to assume that this data came from a truncated exponential population.

55.   (a)

Sample size, n	z value	P-value (upper)
100	$\dfrac{(\bar{x}-\mu_0)}{\sigma/\sqrt{n}} = \dfrac{(101-100)}{15/\sqrt{n}} = \sqrt{n}\,/15 = \sqrt{100}\,/15 \approx .67$	.2514
400	$\sqrt{n}\,/15 = \sqrt{400}\,/15 \approx 1.33$	.0918
1600	$\sqrt{n}\,/15 = \sqrt{1600}\,/15 \approx 2.67$	.0038
2500	$\sqrt{n}\,/15 = \sqrt{2500}\,/15 \approx 3.33$	.0004

(b) $\beta = P(\text{Type II error}) = P(\text{don't reject } H_0 \text{ when } H_0 \text{ is false}) = P(\bar{x} < 100+(2.33)(15)/$

$\sqrt{n}$), when $\mu = 101) = P(z < (100+(2.33)(15)/\sqrt{n} - 101)/(15/\sqrt{n})) = P(z < 2.33 - \sqrt{n}\,/15)$.

Sample size	β
100	$P(z < 2.33 - \sqrt{100}\,/15) = P(z < 1.66) = .9515$
400	$P(z < 2.33 - \sqrt{400}\,/15) = P(z < 1.00) = .8413$
1600	$P(z < 2.33 - \sqrt{1600}\,/15) = P(z < -.33) = .3707$
2500	$P(z < 2.33 - \sqrt{2500}\,/15) = P(z < -1.00) = .1587$

Because the value $\mu = 101$ is so close to the null hypothesis value $\mu = 100$, all the β values above are fairly large.

57.   A 2-sided hypothesis test of $H_0:\mu = 100$ versus $H_a: \mu \neq 100$ is to be conducted using a sample of size n = 15 and a significance level of .01. The researchers believe that the population standard deviation σ is somewhere between .8 and 1 and they want to determine the power of the test for detecting deviations from the null hypothesis value of 100 of up to $.5\sigma$ and $.8\sigma$.

First printout (with $\sigma = 1$):

19.44% of the time, the test will be able to detect a $.5\sigma$ (i.e., .5 pCi/L) deviation from $\mu = 100$; 56.19% of the time it will detect a $.8\sigma$ (i.e., .8 pCi/L) shift away from $\mu = 100$.

128

Second printout (with $\sigma = .8$):

33.44% of the time, the test will be able to detect a $.5\sigma$ (i.e., $.5(.8) = .4$ pCi/L) deviation from $\mu = 100$; 79.67% of the time it will detect a $.8\sigma$ (i.e., $.8(.8) = .64$ pCi/L) shift away from $\mu = 100$.

Third printout (with $\sigma = .8$):

Since the power in the preceding two printouts is very small, the researchers also computed the minimum necessary sample sizes needed to achieve a power of 90% of detecting $.5\sigma$ and $.8\sigma$ shifts from the hypothesized mean of 100. These sample sizes are 42 and 19, respectively.

59.   Combining both samples into one group and ranking:

Observation: 179 183 216 229 232 245 247 250 286 299
Sample #:   2   2   2   1   2   1   2   1   1   1
Rank: 1   2   3   4   5   6   7   8   9   10

The Wilcoxon test statistic is w = sum of ranks for first sample = $4+6+8+9+10 = 37$. The total possible sum of ranks in a sample of size $n_1 = 5$ varies between a minimum of $1+2+3+4+5 = 15$ and a maximum of $6+7+8+9+10 = 40$. Because this is a 2-sided test, we include values of w that are as close to the minimum as the value of 37 is to the maximum possible sum. Therefore, the P-value $= P(w \geq 37) + P(w \leq 18) = .028 + .028 = .056$. For the hypotheses $H_0: \mu_1 - \mu_2 = 0$ versus $H_a: \mu_1 - \mu_2 = 0$, we conclude at significance level, say $\alpha = .05$, that H_0 should not be rejected. There is not sufficient evidence to conclude that the two average bond strengths are different.

61.   (a) Let μ denote the true average soil heat flux covered with coal dust. The sample mean and standard deviation of this data are $\bar{x} = 30.79$ and $s = 6.530$. The relevant hypotheses are $H_0: \mu = 29.0$ versus $H_a: \mu > 29.0$ and the calculated test statistic value is

129

$$t = \frac{\bar{x} - \mu_0}{s/\sqrt{n}} = \frac{30.79 - 29.0}{6.530/\sqrt{8}} = .775. \text{ For d.f.} = n\text{-}1 = 8\text{-}1 = 7, \text{ the P-value} \approx .225$$

(from Table VI), so we would not reject H_0 at significance level $\alpha = .05$. Therefore it can not be concluded that coal dust increases the true average heat flux.

(b) A P-value > .10 for the Ryan-Joiner test indicates that the hypothesis H_0:*population is normal* can not be rejected. This is good, since normality of the population is one of the requirements for the validity o the t-test in part (a). So this result does support the use of the t-test in part (a).

63.   A normal probability plot for this data appears below. Refer to the answer to Exercise 43 of Chapter 2 for a reminder on how to construct a quantile plot in Minitab. The graph approximately follows a straight line, with no obvious outliers, so it is reasonable to assume that the data came from a normal population.

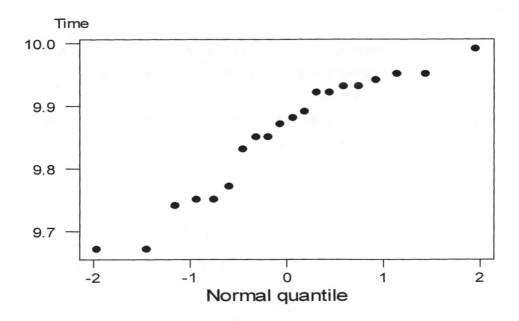

The mandated flame time of 9.75 is a *maximum* standard, so the relevant hypotheses to test are $H_0: \mu = 9.75$ versus $H_a: \mu > 9.75$ (i.e., we want to see if the data shows flame times that exceed the allowed maximum). To conduct the one-sample t test, we first compute the sample mean and standard deviation $\bar{x} = 9.8525$ and $s = .0965$ and the test statistic

$$t = \frac{\bar{x} - \mu_0}{s/\sqrt{n}} = \frac{9.8525 - 9.75}{.0965/\sqrt{20}} = 4.75$$

Using table VI with d.f. $= n-1 = 20 - 1 = 19$, the value 4.75 is larger than any entry in the d.f. $= 19$ column of the table, so the P-value $\approx .000$. Therefore, at the usual significance levels (e.g., .01, .05), H_0 is rejected and we conclude that the average flame time exceeds the allowable maximum of 9.75.

65.   (a) The uniformity specification is that σ not exceed .5, so the relevant hypotheses are $H_0: \sigma = .5$ versus $H_a: \sigma > .5$ (i.e, we want to see if the data shows that the specified uniformity has been exceeded). The test statistic is $\chi^2 = (n-1)s^2/\sigma^2 = (10-1)(.58)^2/(.5)^2 = 12.1104$. From the χ^2 table (Table VII) with d.f. $= n-1 = 10-1 = 9$, we note that 12.1104 is smaller than the smallest entry (14.68) in the d.f. $= 9$ column, so the P-value $> .10$. Therefore, H_0 should not be rejected at any of the usual significance levels (e.g., .05, .01) and we conclude that the data does not contradict the uniformity specification.

        (b) To test $H_0: \sigma = .70$ versus $H_a: \sigma > .70$, the test statistic would be $\chi^2 = (n-1)s^2/\sigma^2 = (10-1)(.58)^2/(.7)^2 = 6.1787$. As in part (a), the exact P-value is not contained in the book's tables, but we can conclude that the P-value must exceed .10.

67.   Let π denote the proportion of students in the population who have cheated in this manner. The relevant hypotheses are then $H_0: \pi = .20$ versus $H_a: \pi > .20$ (i.e., we want to know if the data shows that the proportion is even greater than 20%). The sample proportion is $p = x/n = 124/480 = .2583$, so the test statistic is:

131

$z = \dfrac{p - \pi_0}{\sqrt{\frac{\pi_0(1-\pi_0)}{n}}} = \dfrac{.2583 - .20}{\sqrt{\frac{(.20)(.80)}{480}}} \approx 3.19$. From Table I, the upper-tailed P-value associated

with 3.19 is P-value = $P(z > 3.19) = P(z < -3.19) = .0007$. With a P-value this small, H_0 would certainly be rejected at the usual significance levels (e.g., .05, .01), so we conclude that this is strong evidence that more than 20% of all students have cheated in the manner described in the article.

69.   Let μ_1 denote the true average tear length for Brand A and let μ_2 denote the true average tear length for Brand B. The relevant hypotheses are $H_0: \mu_1-\mu_2 = 0$ versus $H_a: \mu_1-\mu_2 > 0$ (i.e., does the data show that the tear strength is greater for Brand A than Brand B?). Assuming both populations have normal distributions, the 2-sample t test is appropriate. Here, $n_1 = 16$, $\bar{x}_1 = 74.0$, $s_1 = 14.8$, $n_2 = 14$, $\bar{x}_2 = 61.0$, and $s_2 = 12.5$, so the approximate d.f. is:

$$\text{d.f.} = \dfrac{\left[\frac{14.8^2}{16} + \frac{12.5^2}{14}\right]^2}{\left[\frac{14.8^2}{16}\right]^2 \Big/ 15 + \left[\frac{12.5^2}{14}\right]^2 \Big/ 13} = 27.97, \text{ which we round \underline{down} to d.f.} = 27. \text{ The test statistic}$$

is: $t = \dfrac{\bar{x}_1 - \bar{x}_2}{\sqrt{\frac{s_1^2}{n_1} + \frac{s_1^2}{n_2}}} = \dfrac{74.0 - 61.0}{\sqrt{\frac{14.8^2}{16} + \frac{12.5^2}{14}}} \approx 2.6$. From Table VI, the upper-tail P-value associated

with t = 2.6 is P-value = $P(t > 2.6) = .007$. At a significance level of $\alpha = .05$, H_0 is rejected and we conclude that the average tear length for Brand A is larger than for Brand B.

71.   (a) The relevant hypotheses are $H_0: \mu_1^*-\mu_2^* = 0$ (which is equivalent to saying $\mu_1 -\mu_2 = 0$) versus $H_a: \mu_1^*-\mu_2^* \neq 0$ (which is the same as saying $\mu_1 -\mu_2 \neq 0$). The pooled t test is based on d.f. = $n_1+n_2-2 = 8+9-2 = 15$. The pooled variance is:

$$s_p^2 = \dfrac{(n_1-1)s_1^2 + (n_2-1)s_2^2}{n_1 + n_2 - 2} = \dfrac{(8-1)(4.9)^2 + (9-1)(4.6)^2}{8+9-2} = 22.49, \text{ so } s_p = 4.742.$$

The test statistic is then $t = \dfrac{\bar{x}_1^* - \bar{x}_2^*}{s_p\sqrt{\frac{1}{n_1}+\frac{1}{n_2}}} = \dfrac{18.0 - 11.0}{4.742\sqrt{\frac{1}{8}+\frac{1}{9}}} = 3.04 \approx 3.0.$ From Table VI

with d.f. = 15, the P-value associated with $t = 3.0$ is P-value $= 2P(t > 3.0) = 2(.004) =$
.008. At significance level $\alpha = .05$, H_0 is rejected and we conclude that there is a
difference between μ_1^* and μ_2^*, which is equivalent to saying that there is a difference
between μ_1 and μ_2.

(b) No. The mean of a lognormal distribution is $\mu = e^{\mu^* + (\sigma^*)^2/2}$, where μ^* and σ^* are the
parameters of the lognormal distribution (i.e. the mean and standard deviation of ln(x)).
So, when $\sigma_1^* = \sigma_2^*$, then $\mu_1^* = \mu_2^*$ would indeed imply that the $\mu_1 = \mu_2$. However, when
$\sigma_1^* \neq \sigma_2^*$, then even if $\mu_1^* = \mu_2^*$, the two means μ_1 and μ_2 (given by the formula above)
would not be equal.

73.   This is paired data, so the paired t test is employed. The relevant hypotheses are $H_0\!:\!\mu_d = 0$
versus $H_a\!:\!\mu_d < 0$, where μ_d denotes the difference between the population average control
strength minus the population average heated strength. The observed differences (control
minus heated) are: -.06, .01, -.02, 0, and -.05. The sample mean and standard deviation of
the differences are $\bar{x}_d = -.024$ and $s_d = .0305$. The test statistic is:

$t = \dfrac{\bar{x}_d}{s/\sqrt{n}} = \dfrac{-.024}{.0305/\sqrt{5}} = -1.76 \approx -1.8.$ From Table VI, with d.f. = n-1 = 5-1 = 4, the lower-

tailed P-value associated with $t = -1.8$ is P-value $= P(t < -1.8) = P(t > 1.8) = .073.$ At a
significance level of $\alpha = .05$, H_0 should not be rejected. Therefore, this data does not show
that the heated treated average strength exceeds the average strength for the control
population.

75.   The relevant hypotheses are H_0:*the leader's winning % is homogeneous across all 4 sports*
versus H_a:*the leader's winning % differs among the 4 sports*. The appropriate test is a χ^2 test
of homogeneity of several propportions. The following table shows the actual observations
along with the expected values (underneath each observation) for the χ^2 test:

	Leader wins	Leader loses	totals
Basket ball	150 / 155.28	39 / 33.72	189
Baseball	86 / 75.59	6 / 16.41	92
Hockey	65 / 65.73	15 / 14.27	80
Football	72 / 76.41	21 / 16.59	93
Totals	373	81	454

The test statistic value is:

$$\chi^2 = \begin{array}{l} 0.180 + 0.827 + \\ 1.435 + 6.607 + \\ 0.008 + 0.037 + \\ 0.254 + 1.171 = 10.518 \end{array}$$

From Table VII, with d.f. = (2-1)(4-1) = 3, the P-value associated with $\chi^2 = 10.518$ is P-value $\approx .015$. Since the P-value of .015 is smaller than the specified significance level of $\alpha = .05$, H_0 is rejected and we conclude that the leader's winning % is not the same across all 4 sports. In particular, the win percentages are 79.4% (basketball), 93.5% (baseball), 81.3% (hockey), and 77.4% (football), so it appears that the leader's winning percentage is much higher for baseball than for the other three sports.

77.   The samples are paired since each physician's retrieval time is measured twice, once using slides and once using a computer database. The mean and standard deviation of the 15 differences (difference = 'slide time' minus 'computer time') were $\overline{d} = 20.5385$ and $s_d = 11.9626$. The hypotheses to be tested are H_0: $\mu_{1slide} - \mu_{computer} \leq 10$ and H_a: $\mu_{1slide} - \mu_{computer} > 10$. The calculated test statistic is:

$$t = \frac{\overline{d} - 10}{s_d / \sqrt{n}} = \frac{\overline{d} - 10}{s_d / \sqrt{n}} = \frac{20.5385 - 10}{11.9626 / \sqrt{13}} \approx 3.176$$

Using a significance level of $\alpha = .05$, the critical $t_{.05}$ (df = n-1 = 13-1 = 12) = 1.782. Since t = 3.176 exceeds 1.782, we can conclude that the difference in mean retrieval times does exceed 10 seconds.

79.   The means and standard deviations of the two samples are:

	$\bar{x}$	s
Nitride:	12,428.571	7807.201
Thermal:	53,000.000	38,270.96

Next, since $V(a_1 y_1 + a_2 y_2) = a_1^2 V(y_1) + a_2^2 V(y_2)$ holds for any constants a_1 and a_2 and any independent random variables, we can substitute $a_1 = 4$, $a_2 = -1$, $y_1 = \bar{x}_1$ and $y_2 = \bar{x}_2$ to find:

$$V(4\bar{x}_1 - \bar{x}_2) = 4^2 V(\bar{x}_1) + 1^2 V(\bar{x}_2) = 16(\frac{\sigma_1^2}{n_1}) + 1(\frac{\sigma_2^2}{n_2}) \approx 16(\frac{s_1^2}{n_1}) + 1(\frac{s_2^2}{n_2}) =$$

$16(7807.201)^2/7 + 1(38,270.96)^2/7$, so the standard error of $4\bar{x}_1 - \bar{x}_2$ equals the square root of $V(4\bar{x}_1 - \bar{x}_2)$, or approximately 18,669.703. The test statistic is then:

$$z = \frac{4\bar{x}_1 - \bar{x}_2}{\sqrt{16\frac{s_1^2}{n_1} + 1\frac{s_2^2}{n_2}}} = [4(12,428.571) - 53,000.000] / 18669.703 \approx -.176.$$

The hypotheses being tested are H_0: $4\mu_1 - \mu_2 \geq 0$ versus H_a: $4\mu_1 - \mu_2 < 0$, so for $\alpha = .05$, the critical value is $-z_{.05} = -1.645$. Since z = -1.76 is not smaller than −1.645, we cannot reject H_0 and, therefore, we cannot conclude that the mean of the Thermal population data is more than 4 times that of the Nitride population.

Chapter 9
The Analysis of Variance

1. (a) H_0: $\mu_A = \mu_B = \mu_C$; where μ_i = average strength of wood of Type i.

    (b) When a null hypothesis is <u>not</u> rejected in an ANOVA test it can often be good news. When there is no significant difference between two populations (types of beams in this case), you are then free to use <u>other</u> factors when making decisions about the populations. For instance, in this exercise, the choice is narrowed to Type A or B (both of which were shown by the ANOVA test to be superior in strength to Type C). To make the final decision (between A and B), we might use another factor, such as the *cost* of each type of beam, as a factor in deciding between them (since the ANOVA test shows no significant difference in their strengths).

    (c) This is similar to part (b), except now there is no significant difference between the strengths of any of the three beams, so another factor (e.g., cost) can be used to make a decision about which one to use.

3. The two ANOVA tests will give *identical* conclusions. The reason for this is that an ANOVA test is based on comparing <u>variances</u>, which will not be affected by a calibration error. The calibration will certainly cause the mean of the measurements to shift, but the *variation* of the measurements around the mean will be the same as the variation of accurate measurements around the true mean. For example, if x_i denotes the *true* measured strength of an alloy sample, then $x_i + 2.5$ would be the reading given by the instrument. Letting $\bar{x}$ denote the mean of the true measurements, then $\bar{x} + 2.5$ would be the mean of the instrument's measurements. Because $(x_i - \bar{x}) = (x_i + 2.5) - (\bar{x} + 2.5)$, the sample *variances* of the true measurements and the instrument's measurements will be identical.

5.	Using this method, there is no way of knowing whether or not there is a statistically significant difference between the means (because the within-samples variation is not used in the 'pick the winner' strategy). Consider what happens when there is actually *no* significant difference between the means (assuming an ANOVA test was conducted). This information would not be available from the pick the winner' strategy so, unlike the answer to Exercise 1 above, you would not know that you were free to use *other* factors to choose between the populations based on other criteria (e.g., cost, time, etc.).

7.	We will use a subscript to denote the upper-tail area of the F curve (e.g., $F_{.05}$ denotes the F value that is exceeded by the upper 5% of the F distribution). Then, as shown in Exercise 6, $F_{.05}$ ($df_1= 5$, $df_2 = 8$) $\neq$ $F_{.05}$ ($df_1= 8$, $df_2 = 5$) and $F_{.01}(df_1 =5, df_2= 8)$ $\neq$ $F_{.01}$ ($df_1= 8, df_2= 5$). The point is that the critical values of the F distributions depend upon the *order* as well as the numerical values of df_1 and df_2. Mistakenly reversing the order of df_1 and df_2 will give incorrect F-critical values, which may lead to incorrect ANOVA test conclusions.

9.	Using $\alpha = .05$, $df_1=3$, and $df_2 = 20$, a critical 5% upper-tail value of $F = 3.10$ is found in Table VIII. Since the calculated $F = 4.12$, which exceed the critical value of 3.10, H_0 is rejected and we conclude that is a difference among the four means.

11.	There are $k = 4$ populations (brands) and a total of $n = (5)(4) = 20$ sample observations. Therefore, SST has $n-1 = 19$ d.f., SSTr has $k-1 = 4-1 = 3$ d.f., and SSE has $n-k = 20-4 = 16$ d.f. Since $14{,}713.69 = MSE = SSE/(n-k) = SSE/16$, we solve to find $SSE = (16)(14{,}713.69) = 235{,}419.04$. Next, $SSTr = SST-SSE = 310{,}500.76 - 235{,}419.04 = 75{,}081.72$. Finally, $MSTr = SSTr/(k-1) = 75{,}081.72/3 = 25{,}027.24$ and $F = MSTr/MSE = 25{,}027.24/14{,}713.69 = 1.7009$.

Source	df	SS	MS	F
Brand	3	75,081.72	25,027.24	1.7009
Error	16	235,419.04	14,713.69	
Total	19	310,500.76		

13. (a) Let μ_A, μ_B, and μ_C denote the true (i.e., population) moduli of elasticity of the three grades of lumber. Then the hypotheses of interest are $H_0: \mu_A = \mu_B = \mu_C$ versus H_a: *at least two of the population means are not equal.*

(b) Since the sample sizes are equal, we can simply average the three sample means to find the grand mean $\overline{\overline{x}} = (1.63+1.56+1.42)/3 = 1.5367$. Next, $SSE = (n_1-1)s_1^2 + (n_2-1)s_2^2 + (n_3-1)s_3^2 = (10-1)(.27)^2 + (10-1)(.24)^2 + (10-1)(.26)^2 = 1.7829$ and $SSTr = n_1(\overline{x}_1 - \overline{\overline{x}})^2 + n_2(\overline{x}_2 - \overline{\overline{x}})^2 + n_3(\overline{x}_3 - \overline{\overline{x}})^2 = 10(1.63-1.5367)^2 + 10(1.56-1.5367)^2 + 10(1.42-1.5367)^2 = .22867$. In addition, $k = 3$ and $n = (3)(10) = 30$, so SST has d.f. $= n - 1 = 30-1 = 29$, SSTR has d.f. $= k-1 = 3- 1 = 2$, and SSE has d.f. $- n-k = 30-3 = 27$. Therefore, $MSTr = SSTr/(k-1) = .22867/2 = .11433$ and $MSE = SSE/(n-k) = 1.7829/27 = .066033$. The test statistic is $F = MSTr/MSE = .11433/.066033 = 1.731$.

Since the value of $F = 1.731$ is smaller than those listed in Table VIII (for $df_1 = 2$, $df_2 = 27$), we can conclude that the P-value $> .10$. Since the P-value exceeds $\alpha = .01$, H_0 can not be rejected and we not conclude that the data shows any differences between the three population means.

15. The relevant hypotheses are $H_0: \mu_{L1} = \mu_{L2} = \mu_{L3} = \mu_L$ versus H_a: *at least two of the population means differ.* The ANOVA table from a Minitab printout appears below (*note: we entered the data into columns c1, c1, c3, and c4 and used the 'unstack' ANOVA command*).

```
Analysis of Variance
Source    DF       SS        MS       F        P
Factor     3    1.0559    0.3520    3.96    0.053
Error      8    0.7114    0.0889
Total     11    1.7673
```

Although Minitab computes the exact P-value, we can still approximate it by using $df_1 = 3$, and $df_2 = 8$ in Table VIII. From Table VIII we find the P-value associated with $F = 3.96$ lies somewhere between .05 and .10; i.e., $.05 < $ P-value $ < .10$. At a significance level of $\alpha = .05$, this P-value is not small enough to reject H_0, so we can not conclude that there is difference between the mean percentage of methyl alcohol reported by the four laboratories.

17.   (a) Changing units of measurement amounts to simply multiplying each observation by an appropriate conversion constant, c. In this exercise, c = 2.54. Next, note that replacing each x_i by cx_i causes any sample mean to change from $\bar{\bar{x}}$ to $c\bar{\bar{x}}$ while the grand mean also changes from $\bar{\bar{x}}$ to $c\bar{\bar{x}}$. Therefore, in the formulas for SSTr and SSE, replacing each x_i by cx_i will introduce a factor of c^2. That is, SSTR(for the cx_i data) =

$$n_1(c\bar{x}_1 - c\bar{\bar{x}})^2 + \ldots + n_k(c\bar{x}_k - c\bar{\bar{x}})^2 = n_1 c^2(\bar{x}_1 - \bar{\bar{x}})^2 + \ldots + n_k c^2(\bar{x}_k - \bar{\bar{x}})^2 = (c^2)\text{SSTr}$$

(for the original x_i data). The same thing happened for SSE; i.e., SSE(for the cx_i data) = (c^2)SSE(for the original x_i data). Using these facts, we also see that SST(for the cx_i data) = SSTR(for the cx_i data) + SSE(for the cx_i data) = (c^2)SSTr(for the original x_i data)+ (c^2)SSE(for the original x_i data) = (c^2)[SSTR(for the original x_i data)+SSE(for the original x_i data)] = (c^2)SST(for the original x_i data). The net effect of the conversion, then, is to multiply all the sums of square in the ANOVA table by a factor of $c^2 = (2.54)^2$. Because neither the number of treatments nor the number of observations is altered, the entries in the degrees of freedom column of the ANOVA table is not changed. Notice also that the F-ratio remains unchanged: F(for the cx_i data) = MSTR(for the cx_i data)/MSE(for the cx_i data) = (c^2)MSTR(for the original x_i data)/[$(c)^2$MSE(for the original x_i data)] = MSTR(for the original x_i data)/MSE(for the original x_i data) = F-ratio(for the original x_i data). This makes sense, for otherwise we could change the significance of an ANOVA test by merely changing the units of measurement.

    (b) The argument in (a) holds for *any* conversion factor c, not just for c = 2.54. We can conclude then, that *any* change in the units of measurement will change the 'Sum of Squares' column in the ANOVA table, but the degrees of freedom and F ratio will remained unchanged.

19.   Since the sample sizes are equal, we can simply average the three sample means to find the grand mean $\bar{\bar{x}} = (3.43+3.18+3.22)/3 = 3.2767$. Next, $\text{SSE} = (n_1-1)s_1^2 + (n_2-1)s_2^2 + (n_3-1)s_3^2$

$$= (6-1)(.22)^2 + (6-1)(.13)^2 + (6-1)(.11)^2 = .3870 \text{ and } SSTr = n_1(\bar{x}_1 - \bar{\bar{x}})^2 + n_2(\bar{x}_2 - \bar{\bar{x}})^2 +$$

$n_3(\bar{x}_3 - \bar{\bar{x}})^2 = 6(3.43-3.2767)^2 + 10(3.18-3.2767)^2 + 10(3.22-3.2767)^2 = .21640$. Since $k = 3$ and $n = (3)(6) = 18$, SST has d.f. $= n - 1 = 18-1 = 17$, SSTR has d.f. $= k-1 = 3- 1 = 2$, and SSE has d.f. $- n-k = 18-3 = 15$. Therefore, $MSTr = SSTr/(k-1) = .21640/2 = .10820$ and $MSE = SSE/(n-k) = .3870/15 = .0258$. The test statistic is $F = MSTr/MSE = .10820/.0258 = 4.194$. Since the value of $F = 4.194$ falls between the values 3.68 and 6.36 in Table VIII (for $df_1 = 2$, $df_2 = 15$), we can conclude that $.01 < \text{P-value} < .05$. Since the P-value is smaller than $\alpha = .05$, H_0 is rejected and we conclude that there is a difference between the means for the three age categories.

21.   (a) The relevant hypotheses are H_0: $\mu_0 = \mu_{45} = \mu_{90}$ versus H_a: *at least two of the population means differ.* The ANOVA table from a Minitab printout appears below (*note: we entered the data into columns c1, c1, and c3, and used the 'unstack' ANOVA command*).

```
Analysis of Variance
Source     DF        SS         MS        F         P
Factor      2      25.80      12.90      2.35     0.120
Error      21     115.48       5.50
Total      23     141.28
```

Minitab computes the exact P-value, but we can still approximate it by using $df_1 = 2$ and $df_2 = 21$ in Table VIII. In Table VIII, $F = 2.35$ is smaller than any of the entries in the table (for $df_1 = 2$, $df_2 = 21$), so we know that the P-value $> .10$. Since P-value $> .10 > .05 = \alpha$, H_0 is not rejected and we conclude that there is no evidence of a difference between the mean strengths at the three different orientations.

(b) The results in (a) are favorable for the practice of using wooden pegs because the pegs do not appear to be sensitive to the angle of application of the force. Since the angle at which force is applied could vary widely at construction sites, it is good to know that peg strength is not seriously affected by different construction practices (i.e., possibly different force application angles).

23.   (a) Let x_i denote the true value of an observation. Then $x_i + c$ is the measured value reported by an instrument which is consistently off (i.e., out of calibration) by c units. Therefore, if $\bar{x}$ denotes the mean of the true measurements, then $\bar{x} + c$ is the mean of

the measured values. Similarly, the grand mean of the measured values equals $\bar{\bar{x}} + c$, where $\bar{\bar{x}}$ is the grand mean of the true values. Putting these results in the formula for SSTR, we find SSTr(measured values) $= n_1(\bar{x}_1 + c - (\bar{\bar{x}} + c))^2 + ... + {}_k(\bar{x}_k + c - (\bar{\bar{x}} + c))^2$

$= n_1(\bar{x}_1 - \bar{\bar{x}})^2 + ... + n_k(\bar{x}_k - \bar{\bar{x}})^2 =$ SSTR(true values). Furthermore, note that any sample variance is unchanged by the calibration problem since the deviations from the mean for the measured data are identical to the deviations from the mean for the true values; i.e., $(x_i + 2.5) - (\bar{x} + 2.5) = (x_i - \bar{x})$. Therefore, SSE(measured values) $=$ $(n_1-1)s_1^2 + ... + (n_k-1)s_k^2 =$ SSE(true values). Finally, because SSTR and SSE are unaffected, so too will SST be unaffected by the calibration error since SST = SSTR +SSE. Thus, *none* of the sums of squares are changed by the calibration error. Obviously, the degrees of freedom are unchanged too, so the net result is that there will be no change in the entire ANOVA table.

(b) Calibration error will not change any of the ANOVA table entries and therefore will not affect the results of an ANOVA test. That is, if all data points are shifted (up or down) by the same amount c, the ANOVA entries will not be affected. However, the means of each sample *will* shift by an amount equal to c.

25. An effects plot only shows the sample means. It does not show the within-samples variance, so the variation between groups can not be compared to the within-group variation.

27. The sample sizes for the $k = 5$ brands are equal, $(n_1 = n_2 = n_3 = n_4 = n_5 = 4)$, so the appropriate degrees of freedom for the Studentized range are $k = 5$ and $n-k = n_1+n_2+n_3+n_4+n_5$ $= 20 - 5 = 15$. Using a significance level of $\alpha = .05$, we find $q_{.05}(5,15) = 4.37$ from Table IX, so the threshold value is $T = q_{.05}\sqrt{\frac{MSE}{n_i}} = (4.37)\sqrt{\frac{272.8}{4}} = 36.09$. Arranging the sample means from smallest to largest, the means that are no further than $T = 36.09$ apart are underlined:

437.5   462.0   469.3   512.8   532.1

Two of the brands appear to have significantly higher average coverage areas than the other three.

29. The value of T $= 36.09$ will be the same as in Exercise 27. After arranging the sample means from smallest to largest, the ones that are closer than $T = 36.09$ are underlined:

<u>427.5   462.0</u>   469.3   502.8   532.1

The conclusion is similar to that in Exercise 27, but now only one of the brands (the one with mean 532.1) has significantly higher coverage areas than the three with the smallest means.

31. (a) To conduct Tukey's test we need to find the value of MSE. Substituting the given information into the formula for SSE: $SSE = (n_1-1)s_1^2 + (n_2-1)s_2^2 + (n_3-1)s_3^2 + (n_4-1)s_3^2$ $= (121-1)(.62)^2 + (121-1)(.69)^2 + (121-1)(.61)^2 + (121-1)(.55)^2 = 184.212$. Therefore, $MSE = SSE/(n-k) = 184.212/(484-4) = .383775$. The appropriate degrees of freedom for the Studentized range are $k = 4$ and $n-4 = 4(121)-4 = 484-4 = 480$. From Table IX, $q_{.05}(4,480) \approx 3.63$, so the threshold value is $T = q_{.05} \sqrt{\frac{MSE}{n_i}} = (3.63) \sqrt{\frac{.383775}{121}} = .2044$.

   Arranging the sample means from smallest to largest, the means that are no further than $T = .2044$ apart are underlined:

   <u>-.06   -.03   0   .03</u>

   (b) The conclusion from Tukey's test in part (a) is that there are no significant differences between the true mean measurement errors for the four collection times. The fact that the measurement errors stay relatively the same over the four collection times is good news for GPS receivers, since it means that one does not need to spend more and more time collecting data from the satellites. The smallest collection time (30 seconds) produces data that is not significantly different from data produce by the longest times.

33.   The following ANOVA table was obtained from Minitab (for the data of Exercise 12):

```
Analysis of Variance
Source      DF        SS        MS        F        P
Factor      3       456.50    152.17    17.11    0.000
Error       14      124.50      8.89
Total       17      581.00
```

Here, n = 5+4+4+5 = 18, and k =4, so n-k = 18-4 = 14. From Table IX, the required value of the Studentized range is $q_{.05}(4,14) = 4.11$, so the threshold value for Tukey's test is $T = q_{.05}$

$\sqrt{\frac{MSE}{n_i}} = (4.11) \sqrt{\frac{8.89}{4}} = 6.13$. Note that the smallest of the sample sizes ($n_i = 4$) was used

in the formula for T.   Arranging the sample means from smallest to largest and underling the ones that are closer than T to each other gives:

$\underline{45.5 \quad 50.85} \quad 55.4 \quad 55.28$

35.   (a) The following Minitab ANOVA table was created using the command

MTB> anova c1 = c2 c3 (where the data is in column c1; row and column subscripts are in c2 and c3, respectively):

```
Source      DF        SS        MS        F        P
brand       4       53231     13308     95.57    0.000
level       3       116218    38739    278.20    0.000
Error       12       1671       139
Total       19      171120
```

The F-ratio for '*brand*' is F = 95.57. For $df_1 = 4$ and $df_2 = 12$, the value F = 95.57 has a P-value smaller than .001 (From Table VIII). Since this P-value is smaller than $\alpha = .01$, we can conclude that there is a difference in power consumption among the 5 brands

(b) The F-ratio for *'humidity'* is $F = 278.20$. For $df_1 = 3$ and $df_2 = 12$, the P-value associated with $F = 278.20$ is less than .001 (from Table VIII), so the null hypothesis H_0: *average humidity is the same* is rejected. We conclude that humidity levels do affect power consumption, so it was wise to use humidity as a blocking factor.

37.   (a) The following Minitab ANOVA table was created using the command

MTB> anova c1 = c2 c3 (where the data is in column c1; row and column subscripts are in c2 and c3, respectively)

```
Analysis of Variance for effort

Source      DF         SS         MS         F        P
stool        3     81.194     27.065     22.36    0.000
subject      8     66.500      8.312      6.87    0.000
Error       24     29.056      1.211
Total       35    176.750
```

The F- ratio for *'stool type'* is $F = 22.36$. For $df_1 = 3$ and $df_2 = 24$, the P-value associated with $F = 22.36$ is less than .001 (Table VIII). Since this P-value is less than $\alpha = .05$, we reject H_0:*no difference in average effort among stool types* and conclude that the type of stool does have an effect on effort.

(b) The F-ratio for *'subject'* is $F = 6.87$. Using $df_1 = 8$ and $df_2 = 24$, the P-value associated with $F = 6.87$ is less than .001. Since this P-value is less than $\alpha = 5\%$, we reject H_0:*no difference in average effort between subjects* and conclude that different subjects do show different degrees of rising effort, so using *'subject'* as a blocking factor was a good idea.

39.   (a) The following Minitab ANOVA table was created using the command

MTB> anova c1 = c2 c3 (where the data is in column c1; row and column subscripts are in c2 and c3, respectively):

```
Analysis of Variance for strength

Source      DF          SS          MS        F      P
batch        9       86.793       9.644     7.22   0.000
method       2       23.229      11.614     8.69   0.002
Error       18       24.045       1.336
Total       29      134.067
```

The F- ratio for 'method' is F = 8.69. For $df_1 = 2$ and $df_2 = 18$, the P-value associated with F = 8.69 is between .001 and .01 (Table VIII). Since this P-value is less than $\alpha = .05$, we reject H_0:*no difference in average strength among the different methods* and conclude that curing methods do have differing effects on strength.

(b) The F-ratio for 'batch' is F = 7.22. For $df_1 = 9$ and $df_2 = 18$, the P-value associated with F = 7.22 is less than .001. Since this P-value is less than $\alpha = .05$, we reject H_0:*no difference in average strength between different batches* and conclude that different batches do have an effect on strength.

(c) Ignoring *'batch'* and simply conducting a single factor ANOVA using *'method'*, we obtain the following Minitab output (we used the command MTB> *anova c1 =c2*):

```
Analysis of Variance for strength

Source      DF          SS          MS        F      P
method       2       23.229      11.614     2.83   0.077
Error       27      110.838       4.105
Total       29      134.067
```

The F-ratio for *'method'* is now F = 2.83. For $df_1 = 2$ and $df_2 = 27$, the P-value associated with F = 2.83 is between 5% and 10%, i.e., .05 < P-value < .10. Since this P-value is not smaller than $\alpha = .05$, we would not reject H_0:*no difference in average strength among curing methods* and conclude that curing method does <u>not</u> affect strength. This result is caused by the fact that in this single-factor ANOVA we have ignored differences between batches (whose effects have been incorporated into the MSE which leads to a smaller F-ratio than in part (a). That is, ignoring an important blocking factor ('*batch*') can obscure the differences between levels of the factor you are interested in ('*curing method*').

41.   (a) From Table VIII, $F_{.05}(1,10) = 4.96$ and $t_{.025}(10) = 2.228$ and $(2.228)^2 \approx 4.96$. The equality is approximate because the entries in the F and t table entries are rounded.

      (b) $F_\alpha(df_1=1, df_2) = (t_{\alpha/2})^2$, so for $\alpha = .05$: $F_{.05}(1, df_2) = (t_{05/2})^2 = (t_{.025})^2$, which approaches $(z_{.025})^2 = (1.96)^2 = 3.8416$.

43. $n_1 = n_2 = n_3 = 5$, so $k = 3$ and $n = n_1+n_2+n_3 = 15$. Because the sample sizes are equal, the grand mean is simply the average of the three sample means, $\overline{\overline{x}} = (10+12+20)/3 = 14$. Therefore, $SSTr = n_1(\overline{x}_1 - \overline{\overline{x}})^2 + n_2(\overline{x}_2 - \overline{\overline{x}})^2 + n_3(\overline{x}_3 - \overline{\overline{x}})^2 = 5(10-14)^2 + 5(12-14)^2 + 5(20-14)^2 = 280$. Then, $MSTR = SSTR/(k-1) = 280/(3-1) = 140$ and $MSE = SSE/(n-k) = SSE/(15-3) = SSE/12$, so the F-ratio is $F = MSTR/MSE = 140/MSE = 140/(SSE/12). = 1680/SSE$. For $df_1 = 2$ and $df_2 = 12$, the value of F associated with a right-tail area $\alpha = .05$ is $F = 3.89$. So, to reject H_0 (Condition 1), we must have $F = 1680/SSE > 3.89$ or, $1680/3.89 = 431.88 > SSE$.

      For Condition 2, we first look up $q_{.05}(k, n-k) = q_{.05}(3,12) = 3.77$ in Table IX. Next, $T = q_{.05} \sqrt{\frac{MSE}{n_i}} = (3.77)\sqrt{\frac{SSE/12}{5}}$. Therefore, if none of the three sample means are to be further than T units apart, the *largest* of differences between sample means (i..e, $20 - 10 = 10$) must be smaller than T. That is $10 < T = (3.77)\sqrt{\frac{SSE/12}{5}}$. Solving for SSE, we find $SSE > (10/3.77)^2(60) = 422.16$.

      Therefore, for <u>both</u> Conditions 1 and 2 to hold, we must have $422.16 < SSE < 431.88$.

45.   (a) $n_1 = 9$ and $n_2 = 4$, so $k = 2$ and $n = n_1+n_2 = 9+4 = 13$. The grand mean is the weighted average of the sample means, $\overline{\overline{x}} = [9(-.83)+4(-.70)]/[9+4] = -.79$. $SSE = (n_1-1)s_1^2 + (n_2-1)s_2^2 = (9-1)(.172)^2 + (4-1)(.184)^2 = .33824$ and $SSTr = n_1(\overline{x}_1 - \overline{\overline{x}})^2 + n_2(\overline{x}_2 - \overline{\overline{x}})^2$

$= 9(-.83-(-.79))^2 + 4(-.70-(-.79))^2 = .0468$. Therefore, MSTR = SSTR/(k-1) = .0468/(2-1) = .0468, MSE = SSE/(n-k) = .33824/(13-2) = .03075, and F = MSTR/MSE = .0468/.03075 = 1.522. For $df_1 = 1$ and $df_2 = 11$, the P-value associated with F = 1.522 exceeds .10 (Table VIII). Therefore, since the P-value is larger than $\alpha = .01$, we can not reject H_0:*no difference between average diopter measurements for the symptom 'present' and 'absent'.* The data does not show a significant difference between the two groups of pilots.

(b) The equivalent test from Chapter 8 is the independent samples t-test, <u>assuming equal population variances</u>. Recall that one of the ANOVA assumptions is the k population variances are equal, which is why the 'equal variances' test is used.

(c) First, the sample variances must be pooled to find $s_p^2 = \dfrac{(n_1 - 1)s_1^2 + (n_2 - 1)s_2^2}{n_1 + n_2 - 2} =$

$\dfrac{(9-1)(.172)^2 + (4-1)(.184)^2}{9+4-2} = .33824/11 = .03075$. Note that is exactly the same as the MSE computed in part (a). From this, we find $s_p = .17536$. The test statistic is

$t = \dfrac{\bar{x}_1 - \bar{x}_2}{s_p \sqrt{\frac{1}{n_1} + \frac{1}{n_2}}} = \dfrac{-.83 - (-.70)}{.17536\sqrt{\frac{1}{9} + \frac{1}{4}}} = -1.234$. For df = $n_1 + n_2 - 2 = 9 + 4 - 2 = 11$, the

P-value associated with t = -1.234 (for a 2-sided test) is approximately 2(.128) = .256 (from Table VI). Since the P-value $\approx .256$ exceeds $\alpha = .01$, we do not reject H_0:$\mu_1 - \mu_2 = 0$. That is, the test does not show a significant difference between the two groups of pilots. This the same conclusion as in part (a). Note: It can be shown that the 2-sample test (assuming equal variances) <u>always gives the same conclusion</u> as and ANOVA test with k = 2 populations.

47.   (a) The following Minitab ANOVA table was created using the command

    MTB> anova c1 = c2 c3 (where the data is in column c1; row and column subscripts
    are in c2 and c3, respectively)

```
Analysis of Variance for usage

Source      DF         SS         MS        F       P
air          3       930.1      310.0     2.18   0.143
home         4      4959.7     1239.9     8.73   0.002
Error       12      1705.1      142.1
Total       19      7594.9
```

    (b) Using $df_1 = 3$ and df_2 12, the P-value for F = 2.18 is greater than .10 (Table VIII). Since
    the P-value is larger than $\alpha = .05$, we can not reject H_0:*no difference in usage between
    systems.* That is, the data does not show that there is a difference between the air
    conditioners' electricity usage. Note that the block factor *'home'* is significant at $\alpha = .05$
    (F = 8.73), but this only confirms that different homes do indeed have an effect on
    electricity usage.

Chapter 10
Experimental Design

1.   Replication allows you to obtain an estimate of the *experimental error*, which is sometimes
       thought of as the "noise". That is, we expect a certain amount of natural variation between
       experimental results, even when all factors are held fixed, and we call this variation the
       experimental error). Knowing the magnitude of the experimental error is allows you to know
       when a factor's effect is important or not; important/significant factors are those whose effect
       on the response variable causes changes/variation that is larger in magnitude than the
       experimental error.

3.   (a) and (b): The following surface plot of the function f(x,y) was created in MathCAD. The
       surface is a dome whose maximum point sits over the point x = 2, y = 5 in the x-y plane:

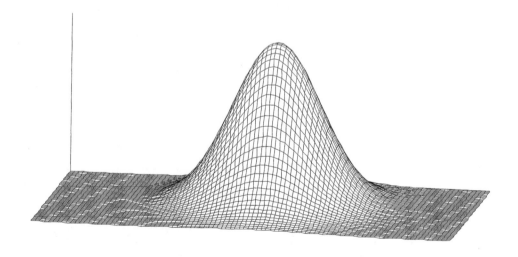

(c) For a contour level of c, we set $f(x,y) = c$. Taking natural logarithms of both sides of the equation $e^{-\frac{1}{2}\left[(x-2)^2+(y-5)^2\right]} = c$, we find $-\frac{1}{2}[(x-2)^2+(y-5)^2] = \ln(c)$. Note that c must be positive (because the exponential function is always positive) and less than 1 (since the expression in the exponent is always negative and the exponential function is less than 1 for negative arguments). Thus, $\ln(c)$ must be negative. Multiplying through by -2 then yields the familiar equation $(x-2)^2+(y-5)^2 = -2\ln(c)$, where $-2\ln(c)$ is a positive number. This is the equation of a circle in the plane with radius equal to the square root of $-2\ln(c)$. That is, all contours for $f(x,y)$ are circles in the plane.

(d) Sketching the contours should show that the maximum is achieved when the expression in the exponent of $f(x,y)$ is 0. Because $(x-2)^2+(y-5)^2$ is nonnegative, it can only equal 0 when both $x = 2$ and $y = 5$. Another method of obtaining the maximum point would be to take partial derivatives of $f(x,y)$, set them equal to 0, and solve. For example, the equations

$$\frac{\partial}{\partial x}\, e^{-\frac{1}{2}\left[(x-2)^2+(y-5)^2\right]} = -(x-2)\, e^{-\frac{1}{2}\left[(x-2)^2+(y-5)^2\right]} = 0 \quad \text{and}$$

$$\frac{\partial}{\partial y}\, e^{-\frac{1}{2}\left[(x-2)^2+(y-5)^2\right]} = -(y-5)\, e^{-\frac{1}{2}\left[(x-2)^2+(y-5)^2\right]} = 0$$

have the unique solution $x = 2$, $y = 5$.

5.   When the lines in the AB interaction plot are parallel, the effect of changing Factor A (or Factor B) from one level to another will be the same for each fixed level of Factor B (or Factor A). That is, when factors A and B do not interact, their effect on the response variable (i.e., the amount that they change the response variable) does not depend on the particular level of the other factor. Since the slope of a line between two values of Factor A is proportional to the *change* in the response, and since this change is not affected by the levels of B, the slope will be the same (i.e., the lines will be parallel) for any fixed level of factor B.

7. Each factor has 5 levels, so each has d.f. = 5-1 = 4. The total number of experimental runs is $n = abr = (5)(5)(3)$, so the total d.f. = abr-1 = 75 - 1 = 74. The AB interaction term has df = (a-1)(b-1) = (5-1)(5-1) = 16. The degrees of freedom for error can then be found by subtracting the factor and interaction degrees of freedom from the total; i.e.; error d.f. = 74 - 4 - 4 - 16 = 50. Alternatively, you could use the formula: error d.f. = ab(r-1) = (5)(5)(3-1) = 50.

Next, MSA = SSA/(a-1) = 20/4 = 5, so the F-ratio for Factor A = MSA/MSE = 5/2 = 2.5. Similarly, the F-ratio for Factor B is 8.1 = F = MSB/MSE = MSB/2, so MSB = 2(8.1) = 16.2 and therefore, 16.2 = MSB = SSB/(b-1) = SSB/4 or, SSB = 4(16.2) = 64.8. Since 2 = MSE = SSE/50, we also have SSE = 2(50) = 100. To find SS(AB), just subtract the SS values for A, B, and error from SST; i.e.; SS(AB) = SST - SSA - SSB - SSE = 200 - 20 - 64.8 - 100 = 15.2. Finally, MS(AB) = SS(AB)/16 = 15.2/16 = .95 and the F-ratio for AB is F = MS(AB)/MSE = .95/2 = .475.

The completed ANOVA table appears below. The point of this exercise is to illustrate that an ANOVA table contains a great deal of redundancy (for the purpose of facilitating making decisions from the table) and to draw attention to the various interrelationships between the entries in the table.

Source	df	SS	MS	F
Factor A	4	20	5	2.5
Factor B	4	64.8	16.2	8.1
Interaction	16	15.2	0.95	0.475
Error	50	100	2	
Total	74	200		

9. (a) Let Factor A be 'formulation' and let Factor B be 'speed'. Then a = 2, b = 3, and the number of replications shown in the data array is r = 3. Therefore, the error d.f. = ab(r-1) = (2)(3)(3-1) = 12, so MSE = SSE/12 = 71.87/12 = 5.9892. MS(AB) = SS(AB)/[(a-1)(b-1)] = 18.58/[(2-1)(3-1)] = 9.29, so the F-ratio for the interaction term is then F = MS(AB)/MSE = 9.29/5.9892 = 1.551. Using df_1 = (a-1)(b-1) = 2 and df_2 = ab(r-1) = 12, the P-value associated with F = 1.551 is greater than .10. Therefore, H_0:*there is no AB interaction effect* can not be rejected. That is, we have no evidence of any interaction between factors A and B.

(b) The F-ratio for Factor A is $F = MSA/MSE = [SSA/(a-1)]/MSE =$ [2253.44/(2-1)]/ 5.9892 = 376.25. Using $df_1 = (a-1) = 1$ and $df_2 = ab(r-1) = 12$, the P-value associated with F = 376.25 is less than .001 (Table VIII), so at significance level $\alpha = .05$ we can reject H_0:*there is no effect for Factor A* and conclude that different chemical formulations do have an effect on yield.

The F-ratio for Factor B is $F = MSB/MSE = [SSB/(b-1)]/MSE = [230.81/(3-1)]/ 5.9892$ = 19.269. Using $df_1 = (b-1) = 2$ and $df_2 = ab(r-1) = 12$, the P-value associated with F = 19.269 is less than .001 (Table VIII), so at significance level $\alpha = .05$ we can reject H_0:*there is no effect for Factor B* and conclude that different speeds also have an effect on yield.

11. (a) The total of all the observations is 27,479. Since each table entry represents the sum of r = 3 replications, the total number of test runs is $n = abr = (3)(5)(3) = 45$, so the grand mean is 27,479/45 = 610.644. Let Factor A be '*capping material*' and let Factor B be '*curing method*'. Then, the column means are: 5432/9, 5684/9, 5619/9, 5567/9, and 5177/9. The row means are: 9410/15, 8835/15, and 9234/15. Therefore,
SSA = $(5)(3)[(9410/15-610.644)^2 + (8835/15-610.644)^2 + (9234/15-610.644)^2] =$ 11,573.38
SSB = $(3)(3)[(5432/9-610.644)^2 + (5684/9-610.644)^2 + (5619/9-610.644)^2 +$ $(5567/9-610.644)^2 + (5177/9-610.644)^2] = 17,930.08$

Finally, SS(AB) = SST - SSA - SSB - SSE = 35,954.31-11,573.38 -17,930.08- 4716.67 = 1734.18. The ANOVA table for this data appears below:

Source	df	SS	MS	F
Material	2	11,573.38	5,786.689	36.81
Batch	4	17,930.08	4,482.521	28.51
Interaction	8	1,734.18	216.773	1.38
Error	30	4,716.67	157.222	
Total	44	35,954.311		

(b) The P-value for F = 36.81 (df$_1$=2, df$_2$=30) is less than .001 (Table VIII), so we reject H$_0$:*no difference in average strengths for the three materials* and conclude that the type of capping material does have an effect on strength. The P-value for F = 28.51 (df$_1$=4, df$_2$=30) is less than .001 (Table VIII), so we reject H$_0$:*no difference in average strength among the five curing methods* and conclude that the type of curing method does have an effect on strength. Finally, the P-value for F = 1.38 (df$_1$=8, df$_2$=30) is greater than .10, so H$_0$:*no interaction between capping material and curing method* can <u>not</u> be rejected; that is, the data does not show that there is any interaction between curing method and capping material.

13. (a) The following ANOVA table was created by *Minitab* using the command:

$$MTB> anova\ c11 \ =c1|c2$$

The wet-mold data is in column c11, and the sand and carbon percentages are in columns c1 and c2; note, Minitab allows only integer 'subscripts', so we used the entries 0, 25 and 50 in column c2. These entries only keep track of the factor levels, so any coding would be acceptable (e.g., we could have used 1, 2 , and 3 for the three levels of carbon fiber percentage instead of 0, 25, and 50).

```
Analysis of Variance for strength
```

Source	DF	SS	MS	F	P
sand	2	705.44	352.72	3.77	0.065
carbon	2	1278.11	639.06	6.83	0.016
sand*carbon	4	278.89	69.72	0.74	0.585
Error	9	842.50	93.61		
Total	17	3104.94			

Exact P-values are included in the ANOVA table and can be used to perform the hypotheses tests. Alternatively, you can use Table VIII:

F = 3.77 (df$_1$ = 2, df$_2$ = 9) → .05 < P-value < .10 → P-value > α = .05, so we do <u>not</u> reject H$_0$:*no effect of sand % on strength*

F = 6.83 (df$_1$ = 2, df$_2$ = 9) → .01 < P-value < .05 → P-value < α = .05, So we <u>do</u> reject H$_0$:*no effect of carbon fiber % on strength*

$F = 0.74$ (df$_1 = 4$, df$_2 = 9$) $\rightarrow$ P-value $> .10$ $\rightarrow$ P-value $> \alpha = .05$,

So we do <u>not</u> reject H$_0$:*no interaction effect between sand % and carbon fiber %*

Therefore, the only thing that appears to affect wet-mold strength is the carbon fiber %.

(b) In a similar fashion, we obtain the following *Minitab* printout for casting hardness:

```
Analysis of Variance for hardness
```

Source	DF	SS	MS	F	P
sand	2	106.778	53.389	6.54	0.018
carbon	2	87.111	43.556	5.33	0.030
sand*carbon	4	8.889	2.222	0.27	0.889
Error	9	73.500	8.167		
Total	17	276.278			

Using $\alpha = .05$, you can use Table VIII or the exact P-values in the ANOVA table to conclude that both '*sand*' and '*carbon*' have a significant effect on hardness, but there appears to be no interaction between the two factors.

(c) For either response variable (wet-mold strength or casting hardness), the results in (a) and (b) show that no interaction is present, so the main effects plots can be consulted for optimum factor settings. The settings that maximize Wet-Mold strength Carbon % = .50% and, since Sand was not a significant factor in Wet-Mold strength, <u>any</u> setting would be acceptable. The settings that maximize Casting Hardness are Sand = 30% and Carbon = 0.20% <u>or</u> 0.50%. Since we eventually have to choose <u>one</u> set of factor levels for the casting process that will optimize <u>both</u> response variables simultaneously, <u>the best choice is Sand % = 30% and Carbon % = .50%</u>.

15. (a) Each factor has three levels: $a = 3$, $b = 3$, $c = 3$. Therefore, the degrees of freedom for each factor is $a-1 = b-1 = c-1 = 2$; the degrees of freedom for each 2-factor interaction is $(a-1)(b-1) = (a-1)(c-1) = (b-1)(c-1) = 4$; and the 3-factor interaction has d.f. = $(a-1)(b-1)(c-1) = 8$. Using these d.f., the sums of squares are converted to mean squares and F-ratios in the ANOVA table:

Source	df	SS	MS	F
A	2	14,144.44	7,072.22	61.06
B	2	5,511.27	2,755.64	23.79
C	2	244,696.39	2,348.20	1,056.27
AB	4	1,069.62	267.20	2.31
AC	4	62.67	15.67	0.14
BC	4	311.67	82.92	0.72
ABC	8	1,080.77	135.10	1.17
Error	27	3,127.50	115.83	
Total	53	270,024.33		

(b) The 2-factor interactions are based on $df_1 = 4$ and $df_2 = 27$. From Table VIII, the P-values of the AB, AC, and BC F-ratios all exceed .10. At a significance level of $\alpha = .05$. none of these P-values are significant; i.e., none of the 2-factor interactions are important.

(c) F-ratios for the main effects (A, B, and C) are based on $df_1 = 2$ and $df_2 = 27$. From Table VIII, the P-values of all three F-ratios are less than .001. Using a significance level of $\alpha = .05$, all three main effects are significant.

17. (a) The numbers of factors levels are: a= 3, b = 2, and c = 4. A Minitab printout of the ANOVA table for this experiment is shown below (*note: In Minitab, put all the data in one column, say c10, and then put the "subcripts" (i.e., factor levels) in columns c1, c2, and c3 and then use the command MTB> anova c10 = c1|c2|c3; the slash marks between columns lets Minitab know to compute the various interaction terms between those columns*)

```
Factor      Type     Levels   Values
   A        fixed       3      1    2    3
   B        fixed       2      1    2
   C        fixed       4      1    2    3    4
```

157

```
Analysis of Variance for length

Source      DF        SS          MS        F        P
A            2      12.896       6.448     1.04    0.360
B            1     100.042     100.042    16.10    0.000
C            3     393.417     131.139    21.10    0.000
A*B          2       1.646       0.823     0.13    0.876
A*C          6      71.021      11.837     1.90    0.092
B*C          3       1.542       0.514     0.08    0.969
A*B*C        6       9.771       1.628     0.26    0.953
Error       72     447.500       6.215
Total       95    1037.833
```

(b) Comparing the P-values (in the ANOVA table) for the 3 main effects (A, B, and C) to $\alpha = .05$, we see that only the main effects for factors B and C have P-values less than .05, so only these effects (i.e., B and C) are important.

(c) None of the F-ratios for the interaction terms are below $\alpha = .05$, so none of the interaction terms are significant.

19. (a) The design matrix is:

A	B	C	response
-	-	-	y_1
+	-	-	y_2
-	+	-	y_3
+	+	-	y_4
-	-	+	y_5
+	-	+	y_6
-	+	+	y_7
+	+	+	y_8

Let y_1, y_2, y_3, ..., y_8 denote the response values (recorded in Yates order). Then, the BC interaction equals $\frac{1}{2}\left[\left(\frac{y_7+y_8}{2}-\frac{y_5+y_6}{2}\right)-\left(\frac{y_3+y_4}{2}-\frac{y_1+y_2}{2}\right)\right]$.

(b) The CB interaction is $\frac{1}{2}\left[\left(\frac{y_7+y_8}{2}-\frac{y_3+y_4}{2}\right)-\left(\frac{y_5+y_6}{2}-\frac{y_1+y_2}{2}\right)\right]$

(c) $BC = \frac{1}{2}\left[\frac{y_7+y_8}{2}-\frac{y_5+y_6}{2}-\frac{y_3+y_4}{2}+\frac{y_1+y_2}{2}\right] = CB$

21. (a) Caution: the data not displayed in Yates order. Let A = PP/LLPDE ratio and let B = RM. Displaying the data in Yates order:

A	B	responses	totals
-	-	10.4, 11.8	22.2
+	-	20.3, 24.5	44.8
-	+	8.1, 9.7	17.8
+	+	19.3, 20.2	39.5

The main effect for A $= $ (A contrast)/$(r2^{k-1})$ = (-22.2+44.8-17.8+39.5)/$(2 \cdot 2^{2-1})$ = 44.3/4 = 11.075.

The main effect for B $=$ (B contrast)/$(r2^{k-1})$ = (-22.2-44.8+17.8+39.5)/$(2 \cdot 2^{2-1})$ = -9.7/4 = -2.425.

The AB interaction = (AB contrast)/$(r2^{k-1})$ = (22.2-44.8-17.8+39.5)/$(2 \cdot 2^{2-1})$ = -.90/4 = -.225.

SSA $=$ (A contrast)$^2/(r2^k)$ = $(44.3)^2/(2 \cdot 2^2)$ = 245.311

SSB $=$ (B contrast)$^2/(r2^k)$ = $(-9.7)^2/(2 \cdot 2^2)$ = 11.761

SS(AB) = (AB contrast)$^2/(r2^k)$ = $(-.90)^2/(2 \cdot 2^2)$ = 0 .101

(b)   Organizing this information into an ANOVA table yields:

Source	df	SS	MS	F
A	1	245.311	245.311	85.44
B	1	11.761	11.761	4.10
AB	1	0.101	0.101	0.04
Error	4	11.485	2.871	
Total	7	268.659		

The P-value for factor A is less than .001 (from Table VIII with $df_1=1$, $df_2 = 4$), but the P-value for factor B exceeds .10. Therefore, only factor A is significant at $\alpha = .05$. Furthermore, then AB interaction term is not significant since the F-ratio of .04 has a P-value ($df_1 = 1$, $df_2 = 4$) that also exceeds .10.

(c)   The only plot that should be interpreted in the main effect for A (since the other effects are not significant. The average response when A is at its high level = 21.075 and the average response when A is at its low level = 10.0, so the plot is:

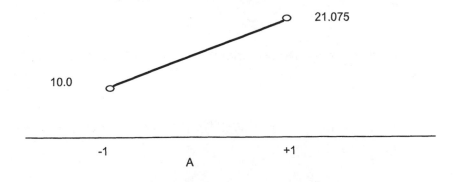

(d)   From the main effect plot in (c), the high level of PP/LLPDE (i.e., factor A) gives the higher response value.   Note that because factor B and the AB interaction are not significant, we can choose *either* level of factor B.

(e) The grand average of the data is $(22.2+44.8+17.8+39.5)/8 = 15.5375$. One-half of the A effect is $11.075/2 = 5.5375$, so the prediction equation is:

$$\hat{y} = 15.5375 + 5.5375x_A.$$

23. (a) Let A = spray volume, B = belt speed, C = brand.

Term	Contrast	Effect = Contrast/(r·2^{k-1})
A	22	$22/(2 \cdot 2^{3-1}) = 2.75$
B	48	$48/(2 \cdot 2^{3-1}) = 6.00$
C	14	$14/(2 \cdot 2^{3-1}) = 1.75$
AB	134	$134/(2 \cdot 2^{3-1}) = 16.75$
AC	-4	$-4/(2 \cdot 2^{3-1}) = -0.50$
BC	-14	$-14/(2 \cdot 2^{3-1}) = -1.75$
ABC	16	$16/(2 \cdot 2^{3-1}) = 2.00$

(b) A Minitab printout of the ANOVA table for this experiment is:

```
Source    DF      SS        MS        F       P
A          1     30.25     30.25     6.72   0.032
B          1    144.00    144.00    32.00   0.000
C          1     12.25     12.25     2.72   0.138
A*B        1   1122.25   1122.25   249.39   0.000
A*C        1      1.00      1.00     0.22   0.650
B*C        1     12.25     12.25     2.72   0.138
A*B*C      1     16.00     16.00     3.56   0.096
Error      8     36.00      4.50
Total     15   1374.00
```

Using the P-values provided in the ANOVA table, only the main effect for B and the AB interaction effect are significant at $\alpha = .01$.

(c) Points on the interaction plot: $A(+)B(-) = 27$, $A(+)B(+) = 49.75$, $A(-)B(-) = 41$,

161

A(-)B(+) = 30.25. A Minitab plot of the B and AB interaction effects follows:

Interaction Plot for surface

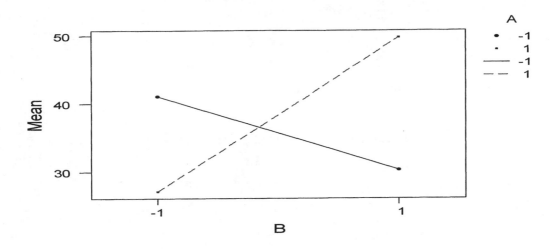

Main Effects for surface

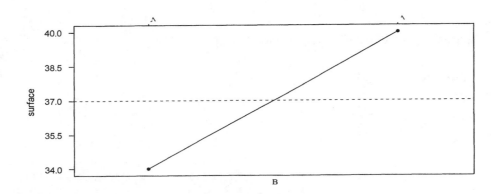

(d) From the AB Interaction plot in (a), both A (spray volume) and B (belt speed) should be set to their high values.

25. (a) Using the DOE option from the Minitab menus, the following effects were calculated:

```
Estimated Effects and Coefficients for combust
Term          Effect        Coef
Constant                    14.312
A             -0.625        -0.313
B              9.625         4.813
C             -4.625        -2.312
A*B            0.175         0.087
A*C            1.125         0.562
B*C           -1.825        -0.913
A*B*C          2.525         1.262
```

(b) A normal probability plot can also be automatically generated within the DOE command in Minitab (by checking the 'normal plot' option). From the Minitab effects plot shown below, Factors B and C appear to be significant.

Normal Probability Plot of the Effects
(response is combust, Alpha = .10)

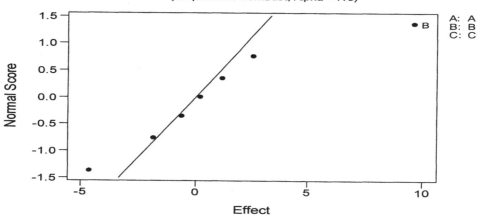

(c) Since the B effect is positive, the high level of B will maximize combustion time. Similarly, the fact that the C effect is negative means that its *low* level will maximize combustion time. Similarly, minimizing combustion time is accomplished by using the low level of B along with the high level of C.

(d) The grand average of the combustion time is 14.313. One-half of the B and C effects
     are 9.625/2 = 4.8125 and -4.625/2 = - 2.3125, respectively. Therefore, the prediction
     model is: $\hat{y} = 14.313 + 4.813x_B - 2.313x_C$

(e) For the response variable 'burnoff', Minitab gives the following effects estimates:

```
Estimated Effects and Coefficients for burnoff
    Term            Effect          Coef
    Constant                        4.371
    A               0.002           0.001
    B              -2.693          -1.346
    C               0.067           0.034
    A*B            -0.132          -0.066
    A*C             0.108           0.054
    B*C             0.063           0.031
    A*B*C          -0.058          -0.029
```

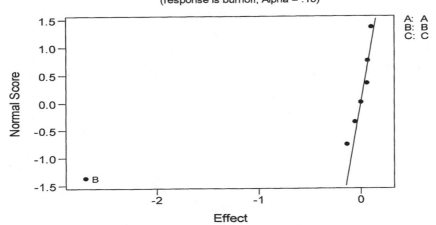

From the effects plot above, only factor B appears to be significant. Because the B effect is
negative, setting B at its low level will maximize the response. Setting B at its high level
will minimize the response. The prediction equation is $\hat{y} = 4.371 - 1.347x_B$.

27.   (a) $2^{k-p} = 2^{7-3}$, so there are k = 7 factors in this design.

(b) The total number of test runs is $2^{7-3} = 2^4 = 16$.

(c) The fraction of the full 2^7 design is $2^{-p} = 2^{-3} = \frac{1}{8}$.

29.   (a) The design matrix is shown below:

A	B	C	D (= AB)
-	-	-	-
+	-	-	+
-	+	-	+
+	+	-	-
-	-	+	+
+	-	+	-
-	+	+	-
+	+	+	+

(b) Multiplication of columns shows that both the AB and CD columns have the same entries:

+

-

-

+

+

-

-

+

Therefore, the AB and CD contrasts will be identical.

31.   (a) There will be $2^{7-2} = 2^5 = 32$ test runs.

   (b) We can express the generators $F = ABCDE$ and $G = CDE$ in the form $ABCDEF = I$ and $CDEG = I$. From these generators, the product of $ABCDEF$ and $CDEG$ is $ABFG$, so the defining relation is then:

$$I = ABCDEF = CDEG = ABFG$$

The alias structure can be summarized as follows:

Main effects are aliased with some 3- and 5-factor interactions;

2-factor interactions are aliased with some 4- and 5-factor interactions;

3-factor interactions are aliased with other 3- and 5-factor interactions.

Note: You can also use the Minitab DOE command to generate the exact alias structure of this design.

33.   (a) There are $k = 5$ factors. Thus, the full 2^k design would have $2^5 = 32$ test runs. Since this design has 16 runs, we conclude that a half-fraction was run. That is $p = 1$.

   (b) Let $A = r_{port}$, $B = h_{lead}$, $C = w_{lead}$, $D = s$, and $E = t_{wall}$. Then, the defining relation is $E = -ABCD$. Notice that, as in this exercise, it is possible to use minus signs when specifying the defining relation. The effect of doing this is simply to specify a *different* fraction of the full 32 runs that would be selected by the relation $E = ABCD$. The rules for multiplying columns still hold, so we find the following alias structure:

Alias Structure:

A	= -BCDE		BC	= -ADE
B	= -ACDE		BD	= -ACE
C	= -ABDE		BE	= -ACD
D	= -ABCE		CD	= -ABE
E	= -ABCD		CE	= -ABD
AB	= -CDE		DE	= -ABC
AC	= -BDE		ABCDE	= I
AD	= -BCE			
AE	= -BCD			

35.   (a) There are k = 4 factors in this experiment. Since the full 2^4 design would require 2^4 = 16 test runs, we see that the 8 runs in this experiment must constitute a half-fraction, so p =1.

(b) Let A = side-to-side, B = yarn type, C = pick density, D = air pressure. The first three columns in the design matrix are in Yates order, but the fourth column (D) is not. After some checking, we find that column D is just the product of the first three columns, so the design generator is D = ABC. The alias structure is then:

A = BCD, B = ACD, C = ABD, D = ABC, AB = CD, AC = BD, AD = BC.

(c) Using the DOE command in Minitab, the effects estimates are:

Estimated Effects and Coefficients for strength

Term	Effect	Coef
Constant		24.3962
A	-0.8975	-0.4487
B	0.4225	0.2112
C	0.7575	0.3788
D	-0.5525	-0.2763
A*B	0.9125	0.4562
A*C	0.0875	0.0438
A*D	-1.0625	-0.5312

Note that all of the alias' are not list in the Minitab printout, so you generally have to fill these in yourself. For example, we know from (b) that AB = CD, so the AB effect of .9125 in the table above is also an estimate of the alias CD, and so forth.

(d) If only effects whose *magnitudes* exceed .7 are important, then (from the table in part (c)), the important effects are: A, C, AB (=CD), and AD (=BC).

(e) A Minitab plot of the main effects (A and C) and the CD and BC interactions are shown below. Points on CD interaction plot: C(+)D(-) = 24.60, C(+)D(+) = 24.96 , C(-)D(-) = 24.75, C(-)D(+) = 23.29. Points on BC interaction plot: B(+)C(-) = 24.76, B(+)C(+) = 24.46, B(-)C(-) = 23.28, B(-)C(+) = 25.10.

Main Effects for strength

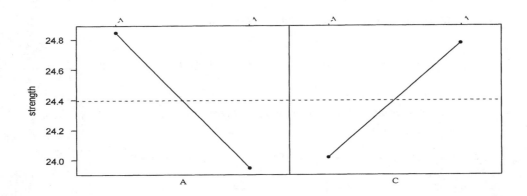

Interaction Plot for strength

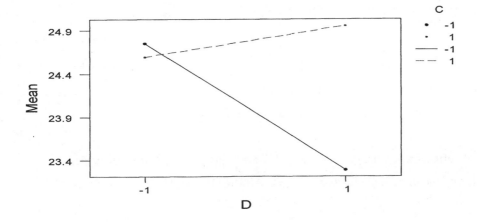

Interaction Plot for strength

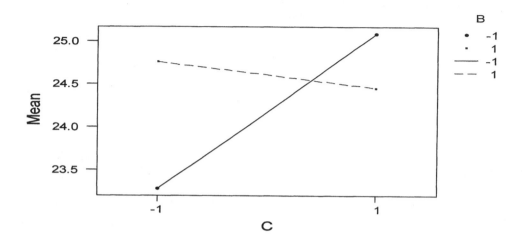

(f)   Now, the only important effects are A and C. from the main effects plots for A and C we see that the settings that maximize fabric strength are: A low and C high.

37.   Only the F-ratios for factors A (F = 16.234) and C (F = 5.348) are significant at α = .01. Main effects plots for factors A and C must be examined to choose settings of these factors; all other factor levels may be chosen arbitrarily.

39.   Caution: test runs are not in Yates order; pooled SS for 2-factor interactions is 18.12 with 10 degrees of freedom, so MSE = 1.812; SSA = 0.856, SSB = 11.391, SSC = 1.380, SSD = 44.56, SSE = 14.25. Factors B, D, and E are significant at α = .01.

Chapter 11
Inferential Methods in Regression and Correlation

1. (a) The slope of the estimated regression line ($\beta = .095$) is the expected <u>change</u> of in the response variable y for each one-unit <u>increase</u> in the x variable. This, of course, is just the usual interpretation of the slope of a straight line. Since x is measured in inches, a one-unit increase in x corresponds to a one-inch increase in pressure drop. Therefore, the expected change in flow rate is .095 m^3/min.

(b) When the pressure drop, x, changes from 10 inches to 15 inches, then a 5 unit increase in x has occurred. Therefore, using the definition of the slope from (a), we expect about a $5(.095) = .475$ m^3/min. increase in flow rate (it is an *increase* since the sign of $\beta = .095$ is *positive*).

(c) For x = 10, $\mu_{y.10} = -.12 + .095(10) = .830$. For x = 15, $\mu_{y.15} = -.12 + .095(15) = 1.305$.

(d) When x = 10, the flow rate y is normally distributed with a mean value of $\mu_{y.10} = .830$ and a standard deviation of $\sigma_{y.10} = \sigma = .025$. Therefore, we standardize and use the z table to find:

$$P(y > .835) = P(z > \tfrac{.835 - .830}{.025}) = P(z > .20) = 1 - P(z \le .20) = 1 - .5793 = .4207 \text{ (using Table I)}.$$

3. (a) The simple linear regression model states that $y = \alpha + \beta x$. Here, $y = \ln(V)$ and $x = 1/T$, so the model becomes $\ln(V) = \alpha + \beta(1/T) + e$. Exponentiating both sides gives: $\exp(\ln(V)) = \exp(\alpha + \beta(1/T) + e)$, or, $V = e^{\alpha} e^{\beta/T} \varepsilon = e^{\alpha} (e^{\beta})^{1/T} \varepsilon = \gamma_0 (\gamma_1)^{1/T} \varepsilon$, where ε is the antilog of the error term e, $\gamma_0 = e^{\alpha}$, and $\gamma_1 = e^{\beta}$. To summarize, the model is

$V = \gamma_0 (\gamma_1)^{1/T} \varepsilon$, which is the model for a multiplicative relationship between the response and predictor variables.

(b) For estimation, we normally set the error term equal to its expected value. For a multiplicative model, the expected value of the error term is 1 (for additive models the expected value is 0). So, for $\alpha = 20.607$, $\beta = -5200.762$ and a temperature of $T = 300$, we predict a vapor pressure of about $V = \gamma_0 (\gamma_1)^{1/T} \varepsilon = (e^{20.607})(e^{-5200.762})^{1/300}(1) = 26.341$.

5.   (a) Using the formulas for the various sums of squares, we find:

$$SS_{xy} = \sum x_i y_i - \frac{1}{n}\left(\sum x_i\right)\left(\sum y_i\right) = 40.968 - (12.6)(27.68)/9 = 2.216$$

$$SS_{xx} = \sum x_i^2 - \frac{1}{n}\left(\sum x_i\right)^2 = 18.24 - (12.6)^2/9 = .600.$$

Therefore, the estimated slope is: $b = SS_{xy}/SS_{xx} = 2.216/.600 = 3.6933$. The estimated intercept is $a = \bar{y} - b\bar{x} = (27.68)/9 - (3.6933)(12.6)/9 = -2.0951$. The estimated regression line is then: $\hat{y} = a + bx = -2.0951 + 3.6933x$.

(b) For $x = 1.5$, the point estimate of the *average* y value is: $\mu_{y\cdot 1.5} \approx -2.0951 + 3.6933(1.5)$ $= 3.445$. For another measurement made when $x = 1.5$, the point estimate of the y value (for this x value) would be the same; i.e., $\hat{y} = 3.445$.

(c) $SSTo = SS_{yy} = \sum y_i^2 - \frac{1}{n}\left(\sum y_i\right)^2 = 93.3448 - (27.68)^2/9 = 8.2134$. Therefore, $SSResid = SSTo - b \cdot SS_{xy} = 8.2134 - (3.6933)(2.216) = .0290$. The estimate of σ is then $s_e \approx \sqrt{\frac{SS\,Resid}{n-2}} = \sqrt{\frac{.0290}{9-2}} = .0644.$

(d) The coefficient of determination is: $r^2 = 1 - \frac{SS\,Resid}{SSTo} = 1 - \frac{.0290}{8.2134} = .996$. Almost all (i.e., about 99.6%) of the observed variation in diffusivity can be attributed to the simple linear regression model between diffusivity and temperature.

7.   (a) Letting y denote the variable 'time', the regression model for the variables y′ and x′ is: $\log_{10}(y) = y' = \alpha + \beta x' + e$. Exponentiating both sides (i.e., taking antilogs of both sides) gives $y = 10^{\alpha + \beta \log(x) + e} = (10^\alpha)(x^\beta)10^e = \gamma_0 \, x^{\gamma_1} \, \varepsilon$; ie, the model is $y = \gamma_0 \, x^{\gamma_1} \, \varepsilon$, where $\gamma_0 = 10^\alpha$ and $\gamma_1 = \beta$. This model is often called a "power function" regression model.

(b) Using the transformed variables x′ and y′, the necessary sums of squares are:
$SS_{x'y'} = 68.640 - (42.4)(21.69)/16 = 11.1615$ and $SS_{x'x'} = 126.34 - (42.4)^2/16 = 13.9800$. Therefore, $b = SS_{x'y'}/SS_{x'x'} = 11.1615/13.9800 = .79839$ and $a = (21.69/16) - (.79839)(42.4/16) = -.76011$. The estimate of γ_1 is then $\gamma_1 = .7984$ and the estimate of γ_0 is $\gamma_0 = 10^a = 10^{-.76011} = .1737$. The estimated power function model is then, $y = .1737 x^{.7984}$. For $x = 300$, the predicted value of y is $.1737(300)^{.7984} = 16.502$, or, about 16.5 seconds.

9.   Let β denote the true average change in runoff (for each $1m^3$ increase in rainfall). To test the hypotheses $H_0: \beta = 0$ versus $H_a: \beta \neq 0$, the calculated t statistic is $t = b/s_b = .82697/.03652 = 22.64$ which (from the printout) has an associated P-value of $P = 0.000$. Therefore, since the P-value is so small, H_0 is rejected and we conclude that there is a useful linear relationship between runoff and rainfall (no surprise here!).

A confidence interval for β is based on $n-2 = 15-2 = 13$ degrees of freedom. The t critical value for, say, 95% confidence is 2.160 (from Table IV), so the interval estimate is:
$b \pm (t \text{ critical}) \cdot s_b = .82697 \pm (2.160)(.03652) = .82697 \pm .07888 = [.748, .906]$
Therefore, we can be confident that the true average change in runoff, for each $1m^3$ increase in rainfall, is somewhere between $.748m^3$ and $.906m^3$.

11.   Let β denote the true average change in flux (for each 1 cm^{-1} increase in inverse foil thickness). To test the hypotheses $H_0: \beta = 0$ versus $H_a: \beta \neq 0$, the calculated t statistic is $t = b/s_b = .26042/.01502 = 17.34$ which (from the printout) has an associated P-value of $P = 0.000$. Therefore, we reject H_0 and conclude that that there is a useful relationship between the x and y variables.

13.	(a)	From the printout in Problem 21 (Chapter 3), the error d.f. = n-2 = 25, so the t-critical value for a 95% confidence interval is 2.060 (Table IV). The confidence interval is then: b ± (t critical)·s_b = .10748 ± (2.060)(.01280) = .10748 ± .02637 = [.081, .134]. Therefore, we estimate with a high degree of confidence that the true average change in strength associated with a 1 GPa increase in modulus of elasticity is between .081 MPa and .134 MPa.

(b)	Letting β denote the true average change in strength (for each 1 GPa increase in modulus of elasticity), the relevant hypotheses to test are H_0:β = .1 versus H_a:β > .1. The test statistic is t = (b-.1)/s_b = (.10748-.1)/(.01280) = .58 ≈ .6, based on n-2 = 25 degrees of freedom. [*Caution: the t-value from the printout in Problem 21 tests the hypothesis H_0:β = 0, so we don't use that t-ratio*]. The P-value for t = .6 is (from Table VI) P = .277. A large P-value such as this would not lead to rejecting H_0, so there is not enough evidence to contradict the prior belief.

15.	Let ρ denote the correlation coefficient for the population from which the sample pairs were selected. The relevant hypotheses to test are H_0:ρ = 0 versus H_a:ρ ≠ 0. The test statistic is:

$$t = \frac{r\sqrt{n-2}}{\sqrt{1-r^2}} = \frac{.449\sqrt{14-2}}{\sqrt{1-(.449)^2}} = 1.741 \approx 1.7,$$ based on n-2 = 12 degrees of freedom.

The two-sided P-value associated with t = 1.7 is 2(.057) = .114. Since this P-value is greater than most reasonable values of α, H_0 should not be rejected. We conclude that there is not sufficient evidence to show that hydrogen content and gas porosity are linearly related.

17.	(a)	$b = \dfrac{SS_{xy}}{SS_x} = \dfrac{2759.6 - \frac{1}{17}(221.1)(193)}{3056.69 - \frac{1}{17}(221.1)^2} = \dfrac{249.4647}{181.0894} = 1.37767 \approx 1.378$ so

there is, on average, an increase of 1.378% in reported nausea for each unit increase in motion sickness dose.

(b) SSResid $=$ SS$_{yy}$ - bSS$_{xy}$ $= [2975 - (193)^2/17] - (1.37767)(249.4647) = 440.273$, so

$s_e^2 =$ SSresid/(n-2) $= 440.273/(17-2) = 29.35153$ and $s_b = \dfrac{s_e}{\sqrt{SS_{xx}}} = \dfrac{5.4174}{\sqrt{181.0894}}$

$= .4026$ and, therefore, $t = \dfrac{b}{s_b} = \dfrac{1.37767}{.4026} = 3.422$. Using a significance level of α

$= .05$, the critical t_α value for a one-sided test of $H_0 : \beta \leq 0$ versus $H_a : \beta > 0$ is

$t_{.05}$ (17-2 = 15 d.f.) $= 1.753$. Since $t = 3.422$ exceeds 1.753, we can conclude that there
is a useful relationship between the two variables.

(c) It would be *possible*, but not advisable to use the linear relationship between the two
variables to predict % nausea when x = 5.0 because the value x = 5.0 is outside the
region where the data was gathered. We don't expect the model to provide good
estimates in this region.

(d) The new sums are: $\Sigma x = 221.1 - 6 = 215.1$,

$\Sigma y = 193 - 2.50 = 190.5$,

$\Sigma xy = 2759.6 - 6(2.50) = 2744.6$

$\Sigma x^2 = 3056.69 - 6^2 = 3020.69$

$\Sigma y^2 = 2975 - (2.50)^2 = 2968.75$

Using these sums, the new slope coefficient is b = 1.42366. Although the slope has
changed, it doesn't appear to be affected very much by the elimination of this point

19. (a) The mean of the x data in Problem 21 (Chapter 3) is $\bar{x} = 45.11$. Since x = 40 is closer to
45.11 than is x = 60, the quantity $(40 - \bar{x})^2$ must be smaller than $(60 - \bar{x})^2$. Therefore,
since these quantities are the only ones that are different in the two $s_{\hat{y}}$ values, the
$s_{\hat{y}}$ value for x = 40 must necessarily be smaller than the $s_{\hat{y}}$ value for x = 60. Said
briefly, the closer x is to $\bar{x}$, the smaller the value of $s_{\hat{y}}$.

(b) From the printout in Problem 21 (Chapter 3), the error degrees of freedom is d.f. = 25. Therefore, for a 95% confidence interval the t-critical value is t = 2.060 (Table IV), so the interval estimate when x = 40 is:

$$\hat{y} \pm (\text{t-critical}) \; s_{\hat{y}} = 7.592 \pm (2.060)(.179) = 7.592 \pm .369 = [7.223, 7.961]$$

We estimate, with a high degree of confidence, that the true average strength for all beams whose MoE is 40 GPa is between 7.223 MPa and 7.961 MPa.

(c) From the printout in Problem 21 (Chapter 3), $s_e = .8657$, so the 95% prediction interval is: $\hat{y} \pm (\text{t-critical}) \sqrt{s_e^2 + s_{\hat{y}}^2} = 7.592 \pm (2.060) \sqrt{(.8657)^2 + (.179)^2} = 7.592 \pm 1.821 =$

[5.771, 9.413]. Note that the prediction interval is almost 5 times (1.821/.369 = 4.93 ≈ 5) as wide as the confidence interval.

(d) For two 95% intervals, the simultaneous confidence level is at least 100(1-2(.05)) = 90%.

21.   n= 15, so the error d.f. = n -2 = 15-2 =13 and, therefore, the t-critical value for a 95% prediction interval is 2.160 (from Table IV). The prediction interval for x = 40 is centered at $\hat{y} = -1.128 + .82697(40) = 31.9508$. The prediction interval is then:

$$\hat{y} \pm (\text{t-critical}) \sqrt{s_e^2 + s_{\hat{y}}^2} = 31.9508 \pm (2.160) \sqrt{(5.24)^2 + (1.44)^2} = 31.95 \pm 11.74$$

= [20.21, 43.69]. Even though the r^2 value is large ($r^2 = 97.5\%$), the prediction interval is rather wide, so precise information about future runoff levels can not be obtained from this model.

23.   (a) Let β denote the true average change in milk protein for each 1 kg/day increase in milk production. The relevant hypotheses to test are $H_0: \beta = 0$ versus $H_a: \beta \pm 0$. The test statistic is $t = b/s_b$ based on n-2 = 14-2 = 12 degrees of freedom. In order to find s_b, we first find s_e:

$s_e^2 = \frac{SS\,Resid}{n-2} = \frac{.02120}{12} = .001767$, so $s_e = .0420$. Then, $s_b = \frac{s_e}{\sqrt{SS_{xx}}} = \frac{.0420}{\sqrt{762.012}} = .00152$,

which then gives a calculated t value of $t = b/s_b = .024576/.00152 \approx 16.2$. The 2-sided P-value associated with $t = 16.2$ is approximately $2(.000) = .000$ (Table VI), so H_0 is rejected in favor of the conclusion that there is a useful linear relationship between protein and production. We should not be surprised by this result since the r^2 value for this data is .956.

(b) For a 99% confidence interval based on d.f. = 12, the t-critical value is 3.055 (from Table IV). The estimated regression line gives a value of $\hat{y} = .175576 + .024576(30) = $

.913 when x = 30. Therefore, $s_{\hat{y}} = (.0420)\sqrt{\frac{1}{14} + \frac{(30-29.56)^2}{762.012}} = .01124$ and the 95% confidence interval is then: $.913 \pm (3.055)(.01124) = .913 \pm .034 = [.879, .947]$.

(c) The 99% prediction interval for protein from a single cow is:

$\hat{y} \pm (t\text{-critical})\sqrt{s_e^2 + s_{\hat{y}}^2} = .913 \pm (3.055)\sqrt{(.0420)^2 + (.01124)^2} = .913 \pm .133$

$= [.780, 1.046]$.

25. (a) The mean value of y, when $x_1 = 50$ and $x_2 = 3$ is $-.800 + .060(50) + .900(3) = 4.9$ hours.

(b) When the number of deliveries (x_2) is held fixed, then average change in travel time associated with a one-mile (i.e., one unit) increase in distance traveled (x_1) is .060 hours. Similarly, when the distance traveled (x_1) is held fixed, then the average change in travel time associated with one extra delivery (i.e., a one-unit increase in x_2) is .900 hours.

(c) Under the assumption that y follows a normal distribution, the mean and standard deviation of this distribution are 4.9 (because $x_1 = 50$ and $x_2 = 3$) and $\sigma = .5$ (since σ is assumed to be constant regardless of the values of x_1 and x_2). Therefore, $P(y \leq 6) = P(z \leq (6-4.9)/.5) = P(z \leq 2.20) = .9861$ (from Table I). That is, in the long run, about 98.6% of all days will result in a travel time of at most 6 hours.

27.   (a) A Minitab graph of the function $y = 220 + 75x - 4x^2$ over the region $2 \leq x \leq 12$ appears
         below:

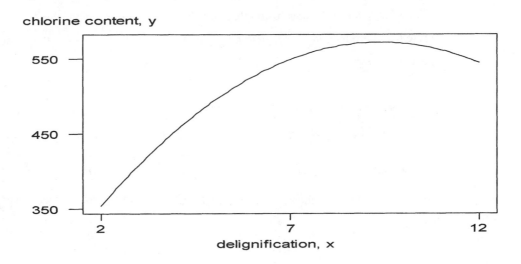

(b) For $x = 8$, $y = 220 + 75(8) - 4(8)^2 = 564$. For $x = 10$, $y = 220 + 75(10) - 4(10)^2 = 570$. So,
     the mean content is slightly higher when the degree of delignification is 10.

(c) For $x = 9$, $y = 220 + 75(9) - 4(9)^2 = 571$. So the change in average chlorine content (y) as
     x changes from 8 to 9 is $571 - 564 = 7$. On the other hand, the same one-unit change in x
     from $x = 9$ to $x = 10$ results in an increase of only $571 - 570 = 1$.

29.   (a) For $x_1 = 2$, $x_2 = 8$ (remember, the units of x_2 are in 1000's), and $x_3 = 1$ (since the outlet
         has a drive-up window), the average sales are $y = 10.00 - 1.2(2) + 6.8(8) + 15.3(1) =$
         77.3 (i.e, $77,300).

(b) For $x_1 = 3$, $x_2 = 5$, and $x_3 = 0$, the average sales are $y = 10.00 - 1.2(3) + 6.8(5) + 15.3(0)$
     $= 40.4$ (i.e, $40,400).

178

31. (a) To test $H_0: \beta_1 = \beta_2 = 0$ versus H_a: *at least one of β_1 and β_2 is not zero*, the test statistic is:

$$F = \frac{MS\,\mathrm{Re}\,gr}{MS\,\mathrm{Re}\,sid} = 319.31 \text{ (from printout).}$$ Alternatively, you could use the formula

$$F = \frac{R^2/k}{(1-R^2)/(n-(k+1))} = \frac{.991/2}{(1-.991)/6} = 330.33.$$ The discrepancy between the two

methods is caused by the rounding in R^2 (e.g., as you can see from the sums of squares in the printout, $R^2 = 715.50/722.22 = .9907$, which was then rounded to .991). Regardless, the P-value associated with and F-value of 319 or so is zero, so, at any reasonable level of significance, H_0 should be rejected. There does appear to be a useful linear relationship between temperature difference and at least one of the two predictors.

(b) The degrees of freedom for SSResid = n-(k+1) = 9 - (2+1) = 6 (which you could simply read in the "DF" column of the printout). Therefore, the t-critical value for a 95% confidence interval is 2.447 (Table IV). The desired confidence interval is:

$b_2 \pm$ (t-critical) s_{b_2} = $3.0000 \pm (2.447)(.4321) = 3.0000 \pm 1.0573$, or about [1.943, 4.057].

That is, holding furnace temperature fixed, we estimate that the average change in temperature difference on the die surface will be somewhere between 1.943 and 4.057.

(c) When x_1 = 1300 and x_2 = 7, the estimated average temperature difference is $\hat{y}$ = -199.56 $+.2100x_1 + 3.000x_2 = -199.56 +.2100(1300) + 3.000(7) = 94.44$. The desired confidence interval is then $94.44 \pm (2.447)(.353) = 94.44 \pm .864$, or, about [93.58, 95.30].

(d) From the printout, s_e = 1.058, so the prediction interval is:

$$94.44 \pm (2.447)\sqrt{(1.058)^2 + (.353)^2} = 94.44 \pm 2.729, \text{ or, about } [91.71, 97.17].$$

179

33.   (a) The appropriate hypotheses are $H_0: \beta_1=\beta_2=\beta_3=\beta_4=0$ versus H_a:*at least one of the β_i's is*

not zero. The test statistic is $F = \dfrac{R^2/k}{(1-R^2)/(n-(k+1))} = \dfrac{.946/4}{(1-.946)/(25-(4+1))} =$

87.6. The test is based on $df_1 = 4$, $df_2 = 20$. From Table VIII, the P-value associated with $F = 6.59$ is .001, so the P-value associated with 87.6 is obviously .000. Therefore, H_0 can be rejected at any reasonable level of significance. We conclude that at least one of the four predictor variables appears to provide useful information about tenacity.

(b) The adjusted R^2 value is $1 - \dfrac{n-1}{n-(k+1)} \dfrac{SS\,Resid}{SSTo} = = 1 - \dfrac{n-1}{n-(k+1)}[1-R^2] =$

$1 - \dfrac{24}{20}[1-.946] = .935$, which does not differ much from $R^2 = .946$.

(c) The estimated average tenacity when $x_1 = 16.5$, $x_2 = 50$, $x_3 = 3$, and $x_4 = 5$ is:

$\hat{y} = 6.121 - .082x_1 + .113x_2 + .256x_3 - .219x_4 = 6.121 - .082(16.5) + .113(50) + .256(3)$

$-.219(5) = 10.091$. For a 99% confidence interval based on 20 d.f., the t-critical value is 2.845. The desired interval is: $10.091 \pm (2.845)(.350) = 10.091 \pm .996$, or, about [9.095, 11.087]. Therefore, when the four predictors are as specified in this problem, the true average tenacity is estimated to be between 9.095 and 11.087.

35.   (a) The negative value of b_2, which is the coefficient of x^2 in the model, indicates that the parabola $b_0 + b_1x + b_2x^2$ opens downward.

(b) $R^2 = 1 - SSResid/Ssto = 1 - .29/202.87 = .9986$, so about 99.86% of the variation in output power can be attributed to the relationship between power and frequency.

(c) With an R^2 this high, it is very likely that the test statistic will be significant. The relevant hypotheses are $H_0: \beta_1=\beta_2=0$ versus H_a:*at least one of the β_i's is not zero.* The

180

test statistic is: $F = \dfrac{SS\,\mathrm{Re}\,gr/k}{SS\,\mathrm{Re}\,sid/(n-(k+1))} = \dfrac{(202.87-2.9)/2}{.29/5} = 1746.$ Clearly, the

P-value associated with $F = 1746$ is 0, so H_0 is rejected and we conclude that the model is useful for predicting power.

(d) The relevant hypotheses are $H_0{:}\beta_2{=}0$ versus $H_a{:}\beta_2 \neq 0$. The test statistic is:

$t = b_2/s_{b_2} = -.00163141/.00003391 = 48.$ The P-value for this statistic is 0 and H_0 is

rejected in favor of the conclusion that the quadratic predictor provides useful information.

(e) The estimated average power when $x = 150$ is $\hat{y} = -1.5127 + .391902x - .00163141x^2$

$= -1.5127 + .391902(150) - .00163141(150)^2 = 20.57.$ The t-critical value based on

5 d.f. is 4.032, so the 99% confidence interval is: $20.57 \pm (4.032)(.1410) = 20.57 \pm .57$ or,

about $[20.00, 21.14]$ To find the prediction interval, we must first find s_e.

$s_e^2 = \mathrm{SSResid}/(n-3) = .29/5 = .058,$ so $s_e = .241.$ Therefore, the 99% prediction interval

is: $20.57 \pm (4.032)\sqrt{(.241)^2 + (.141)^2} = 20.57 \pm 1.13,$ or, about $[19.44, 21.70].$

37. Let $\beta_1,\ \beta_2,\ \beta_3,\ \beta_4,$ and β_5 denote the regression coefficients for $x_1,\ x_2,\ x_1{}^2,\ x_2{}^2,$ and $x_1x_2,$ respectively. We then wish to test $H_0{:}\beta_3{=}\beta_4{=}\beta_5{=}0$ versus $H_a{:}$*at least one of $\beta_3,\ \beta_4,$ and β_5 is not zero.* H_0 asserts that none of the three second-order variables provides useful information. The test statistic is:

$$F = \frac{[\mathrm{SSResid(reduced)} - \mathrm{SSResid(full)}]/g}{\mathrm{SSResid(full)}/[n-(k+1)]} = \frac{[894.95 - 390.64]/3}{390.64/[14-(5+1)]} = 3.44,$$

where g = the number of predictors in the group considered for deletion = 3, k = the number of predictors in the full model = 5, and n = the sample size = 14. The test is based on $df_1 = 3$ and $df_2 = 8$. From the Table VIII we find $.05 < $ P-value $ < .10$. Therefore, at significance level $\alpha = .01$, H_0 cannot be rejected. That is, we do not have evidence that any of the three second-order predictors provide useful information beyond what is already provided by x_1 and x_2 together. Although there seems to be a large difference between the SSResid values for the full and reduced models, the small number of degrees of freedom for error (d.f. error = 8) is simply too small to allow us to conclude that the full model is useful.

181

39.   Both variables are significant at the $\alpha = .05$ level (since both P-values $< .05$), so both
      variables are needed in the model. The $R^2 = 92.3\%$ is very high, which seems to indicate that
      the model fits the data fairly well. However, $s_e = 44.28$ so, roughly, about 95% of the data
      points should lie within 2 standard deviations of the regression line. Since $2(44.28) = 88.56$,
      this isn't a very precise estimate, especially since many of the observed y values are less than
      30.

41.   (a) The following Minitab lot of the standardized residuals versus their corresponding
          normal quantiles shows a reasonable straight line, with the possible exception of the
          point in the upper right corner, $(1.70991, 2.16543)$. It seems reasonable that the
          standardized residuals could have come from a normal distribution.

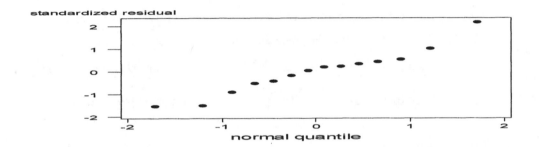

      (b) The separate plots of the standardized residuals versus x_1 (depth) and x_2 (water content)
          are shown below. Neither plot shows any obvious pattern (i.e., non-random behavior),
          so we have no evidence in these plots that the model should be modified.

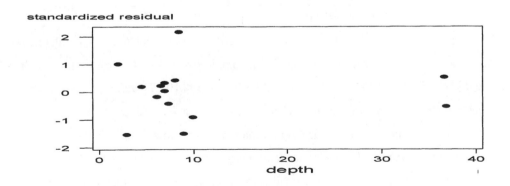

standardized residual

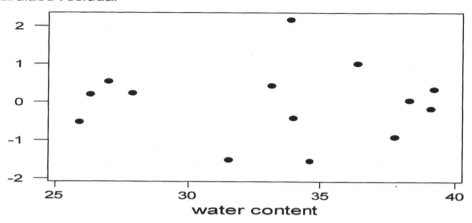

water content

43.   One of the standardized residuals (the first one, 2.01721) slightly exceeds 2, which is not a cause for alarm. To check for the presence of high-leverage observations, we calculate $3(k+1)/n = 3(3+1)/19 = .632$. Because $h_{14,14} = .712933 > .632$, the 14th observation is potentially highly influential. Searching row 14 of the data reveals that the value of $x_3 = .55$ is noticeably different from the other values of x_3 in the remaining columns. The model should be refit with the data in row 14 omitted to see if the calculated quantities (e.g., the b_i's, R^2, etc.) change substantially.

45.   **Backwards stepping**:

Step 1: A model with all 5 variables is fit; the smallest t-ratio is $t = .12$, associated with variable x_2 (summerwood fiber %). Since $t = 1.2 < 2$, the variable x_2 was eliminated.

Step 2: A model with all variables except x_2 was fit. Variable x_4 (springwood light absorption) has the smallest t-ratio ($t = -1.76$), whose magnitude is smaller than 2. Therefore, x_4 is the next variable to be eliminated.

Step 3: A model with variables x_1, x_3, and x_5 is fit. The smallest t-ratio is $t = 1.98$, associated with variable x_1. Since $1.98 < 2$, variable x_1 is the next to be eliminated.

Step 4: A model with variables x_3 and x_5 is fit. Both t-ratios have magnitudes that exceed 2, so both variables are kept and the backwards stepping procedure stops at this step. The final model identified by the backwards stepping method is the one containing x_3 and x_5.

Forwards Stepping:

Step 1: After fitting all 5 <u>one-variable</u> models, the model with x_3 had the t-ratio with the largest magnitude ($t = -4.82$). Because the absolute value of this t-ratio exceeds 2, x_3 was the first variable to enter the model.

Step 2: All 4 <u>two-variable</u> models that include x_3 were fit. That is, the models $\{x_3, x_1\}$, $\{x_3, x_2\}$, $\{x_3, x_4\}$, and $\{x_3, x_5\}$ were fit. Of all 4 models, the t-ratio 2.12 (for variable x_5) was largest in absolute value. Because this t-ratio exceeds 2, x_5 is the next variable to enter the model.

Step 3 (not printed): All possible three-variable models involving x_3, x_5, and another predictor are fit. None of the t-ratios for the added variables (i.e., for x_1, x_2, or x_4) have absolute values that exceed 2, so no more variables are added. There is no need to print anything in this case, so the results of these tests are not shown.

Note: Both the forwards and backwards stepping methods arrived at the same final model $\{x_3, x_5\}$ in this problem. This often happens, but not always. There are cases when the different stepwise methods will arrive at slightly different collections of predictor variables.

47.   If multicollinearity were present, at least one of the four R^2 values would be very close to 1, which is <u>not</u> the case. Therefore, we conclude that multicollinearity is not a problem in this data.

49.   The point estimate of β is $b = .17772$, so the estimate of the odds ratio is $e^b = e^{.17772} \approx 1.194$. That is, when the amount of experience increases by one year (i.e., a one-unit increase in x), we estimate that the odds ratio increase by about 1.194. The z-value of 2.70 and its

corresponding P-value of .007 imply that the null hypothesis $H_0:\beta=0$ can be rejected at any of the usual significance levels (e.g., $\alpha = .10, .05, .025, .01$). Therefore, there is clear evidence that β is not zero, which means that experience does appear to affect the likelihood of successfully performing the task. This is consistent with the confidence interval [1.05, 1.36] for the odds ratio given in the printout,. since this interval does not contain the value 1. A graph of $\hat{\pi}$ appears below.

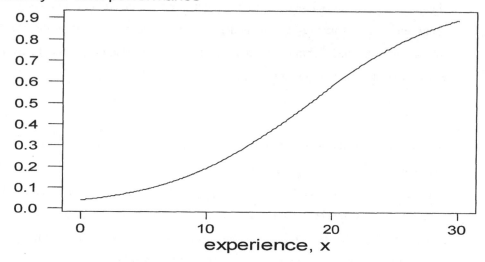

probability of task performance

51.   (a) The five observations made when $x = 500$ all resulted in different values of y. If the relationship were deterministic, all five of these y values would have been identical.

(b) $\sum x_i = 6075$, $\sum y_i = 371$, $\sum x_i y_i = 226{,}565$, $\sum x_i^2 = 3{,}806{,}125$, $\sum y_i^2 = 13{,}861$

so $SS_{xy} = 226{,}565 - \frac{1}{10}(6075)(371) = 1182.5$ and $SS_{xx} = 3{,}806{,}125 - \frac{1}{10}(6075)^2 =$ 115,562.5. Therefore, $b = SS_{xy}/SS_{xx} = 1182.5/115{,}562.5 = .01023256$ and $a = (371/10)$ $- .01023256(6075/10) = 30.883721$. $SST_0 = 13{,}861 - \frac{1}{10}(371)^2 = = 96.90$ and SSResid $= SST_0 - bSS_{xy} = 84.80$. The r^2 value is $r^2 = 1 - SSResid/SST_0 = 1 - 84.80/96.90 = .125$. $s_e^2 = SSResid/(n-2) = 84.80/(10-2) = 10.6$, so $s_e = 3.256$. The standard error of b is $s_b =$

185

$s_e / \sqrt{SS_{xy}} = 3.256 / \sqrt{115,562.5} = .009578$, so the t-ratio is $t = b/s_b = .01023256/.009578$

$= 1.068$, or approximately 1.1. Based on $n-2 = 8$ d.f., the two-sided P-value associated

with $t = 1.1$ is $2(.152) \approx .30$. Since the P-value exceeds any reasonable value of α,

$H_0: \beta = 0$ is <u>not</u> rejected. We can not conclude that the simple linear regression model is

useful.

(c) Because the observations on y were made at only two distinct x values, there is no way to

distinguish between a simple linear regression model (i.e., a straight line) and any

higher-order polynomial model (quadratic, cubic, etc.). For example, to distinguish

between a simple linear regression model and a quadratic model requires that

observations be made for at least <u>three</u> distinct x values so that the potential curvature in

the scatter plot can be observed.

53. (a) To see if the model contain useful information, the F test is performed. The relevant

hypotheses are $H_0: \beta_1 = \beta_2 = \beta_3 = \beta_4 = 0$ versus H_a:*at least one of the β's is not zero* and the

test statistic is $F = \text{MSRegr}/\text{MSResid} = 9.8442/.6998 = 14.07$, with an associated P-value

of .000 (all these numbers come from the printout accompanying the problem). With

such a small P-value H_0 is rejected and we conclude that there is a useful linear

relationship between durable press rating and the four predictors.

(b) This part of the problem asks for a confidence interval for β_3 (the coefficient of curing

temperature). β_3 is the true average change in y (durable press rating) for each one-unit

increase in x_3 (curing temperature), assuming that the other variables (x_1, x_2, and x_4)

remain fixed. From the printout, $b_3 = .011226$ and $s_{b_3} = .004973$. The t-critical value

for a 95% confidence interval based on 25 degrees of freedom (i.e., the error d.f. from

the printout) is 2.060, so the desired interval is:

$b_3 \pm$ (t-critical) $s_{b_3} = .011226 \pm (2.060)(.004973) = .011226 \pm .010244$, or,

approximately [.0010, .0214]. We can be highly confident that $.0010 < \beta_3 < .0214$.

(c) To answer this question we perform the t-test of the hypotheses $H_0:\beta_1 = 0$ versus $H_a:\beta_1 \neq 0$. H_0 can be interpreted as saying that x_1 provides no useful information *beyond the information provided by the remaining three variables (x_2, x_3, and x_4)*. The test statistic is $t = b_1 / s_{b_1} = 2.43$ (from the printout) and corresponding P-value = .023.

Therefore, for a significance level of, say, $\alpha = .05$, H_0 would be rejected and we would conclude that formaldehyde concentration (x_1) does contribute useful information to the model over and above the information provided by the other three variables. Notice, however, that the same can not be said of variable x_4 (curing time) since it has a fairly small t-ratio of 1.74 (with a correspondingly large P-value of .095).

(d) The following printout comes from running Minitab's 'best subsets' regression routine. The predictors x_2, x_2^2, and $x_1 x_3$ appear in a majority of the models. There is no reasonable model that does not depend (in some fashion) on all four predictor variables. For example, although x_1 itself appears in relatively few models, x_1^2 appears in the majority of the models, so the information in x_1 is represented (through x_1^2) in most of the reasonable models. There are a number of defensible choices for the 'best' set of predictor variables. In particular, all three of the best 6-predictor models have an R^2 close to that for the model with all predictors; their adjusted R^2 values are close to the maximum value, and Mallow's CP is reasonably close to $6+1 = 7$. One way of narrowing the choice is to select a likely set of predictors and run the regression model with these predictors. Since all of the partial-t ratios should be large (i.e., significant) in 'good' models, you can discard models that do not have all t-ratios significant.

```
Response is y (durable press rating)
                                           x x x x x x x x x x
                                           1 2 3 4 1 1 1 2 2 3
            R-Sq                   x x x x s s s s x x x x x x
  Vars  R-Sq  (adj)   C-p      S   1 2 3 4 2 q q q 2 3 4 3 4 4

   1    52.5  50.8   64.5  0.98251                         X
   1    52.1  50.4   65.2  0.98614                       X
   2    64.6  62.0   43.4  0.86298  X         X
   2    63.3  60.6   45.9  0.87898  X                 X
   3    78.7  76.2   18.6  0.68304  X         X       X
   3    72.8  69.7   29.8  0.77112  X X       X
   4    82.2  79.4   13.9  0.63607  X     X   X       X
   4    82.1  79.2   14.1  0.63843  X         X   X   X
   5    86.7  83.9    7.4  0.56196  X       X X       X X
   5    85.5  82.5    9.6  0.58571  X       X X X     X
   6    88.5  85.5    5.9  0.53337  X         X X     X X X
   6    88.2  85.1    6.5  0.54094  X       X X X     X           X
   7    89.9  86.7    5.2  0.51045  X X X X X X       X
   7    89.9  86.7    5.3  0.51140  X X       X X X X X
   8    91.5  88.3    4.2  0.47933  X X X X X X       X           X
   8    90.6  87.0    6.0  0.50535  X X       X X X X X           X
   9    91.7  87.9    5.9  0.48650  X X X X X X X     X           X
   9    91.6  87.8    6.0  0.48853  X X X X X X       X     X     X
  10    91.8  87.4    7.7  0.49670  X X X X X X       X     X     X
  10    91.8  87.4    7.7  0.49684  X X X X X X         X X X     X
  11    91.9  87.0    9.3  0.50435  X X X X X X     X X X X       X
  11    91.9  86.9    9.5  0.50682  X X X X X X X       X X X     X
  12    92.0  86.4   11.2  0.51717  X X X X X X X     X X X X     X
  12    92.0  86.3   11.3  0.51815  X X X X X X X X X X X X       X
  13    92.1  85.7   13.1  0.53006  X X X X X X X X X X X X X     X
  13    92.0  85.6   13.2  0.53222  X X X X X X X     X X X X X X
  14    92.1  84.8   15.0  0.54639  X X X X X X X X X X X X X X
```

55. (a) Part of the Minitab printout for the regression of w (life) on s (speed) and l (load) is shown below. The F-test of $H_0: \beta_1 = \beta_2 = 0$ versus H_a: *at least one of β_1 or β_2 is not zero* is conducted using the test statistic value of $F = MSRegr/MSResid = 23.07$ (from the printout), which has a corresponding P-value of .000. Clearly H_0 is rejected (at any reasonable value of α) in favor of the conclusion that there is a useful relationship between y and the two predictor variables. Furthermore, because both predictor variables have large t-ratios (and, therefore, small P-values), it appears that neither predictor should be deleted from the model, even if the other on is retained. Both

predictors appear to provide useful information.

```
The regression equation is
life = 287 - 1.79 speed - 15.9 load

Predictor         Coef         StDev           T          P
Constant        286.81         33.13        8.66      0.000
speed          -1.7863        0.3335       -5.36      0.000
load           -15.867         3.798       -4.18      0.000

S = 56.59        R-Sq = 65.8%      R-Sq(adj) = 62.9%

Analysis of Variance

Source         DF           SS          MS           F          P
Regression      2       147781       73890       23.07      0.000
Error          24        76863        3203
Total          26       224643
```

Despite these results, an analysis of the residuals (see graph below) shows that there is a strong curved pattern in the residuals, which indicates that some action must be taken to correct this problem.

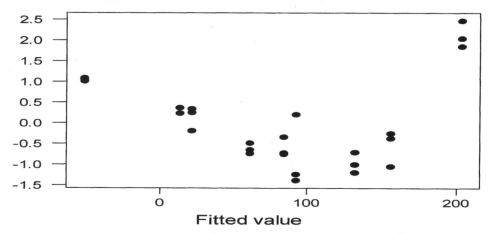

(b)   In an attempt to solve the inadequate model in (a), we instead fit a model of the form $w = [\delta/(s^{\beta}l^{\gamma})]\varepsilon$. By taking natural logarithms of both sides, the model becomes $\ln(w) = \ln(\delta) - \beta\ln(s) - \gamma\ln(l) + \ln(\varepsilon)$ which is of the form $y = \alpha + \beta_1 x_1 + \beta_2 x_2 + e$. That is, we can estimate the parameters of this model by simply performing a

189

regression of ln(w) on the two predictors ln(s) and ln(l). A Minitab printout from such a regression is shown below:

```
The regression equation is
loglife = 10.9 - 1.21 logspeed - 1.40 logload

Predictor          Coef          StDev              T          P
Constant        10.8764         0.7872          13.82      0.000
logspeed        -1.2060         0.1710          -7.05      0.000
logload         -1.3988         0.2327          -6.01      0.000

S = 0.5966        R-Sq = 78.2%        R-Sq(adj) = 76.3%

Analysis of Variance

Source        DF          SS            MS          F          P
Regression     2      30.568        15.284      42.95      0.000
Error         24       8.541         0.356
Total         26      39.109
```

The R^2 value in this printout is substantially higher than the R^2 for the model in part (a). The F-ratio for model utility is $F = 42.95$ (with associated P-value of .000), indicating a useful relationship between the dependent variable and the two predictor variables. Again, both predictor variables have large t-ratios and, therefore, are both needed in the model. Finally, a plot of the residuals for this model reveals no discernible patterns or problems.

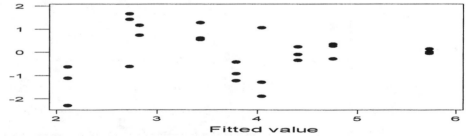

The estimated model is $\ln(w) = 10.8764 - 1.2060\ln(s) - 1.3988\ln(l)$. Taking antilogs of both sides gives the estimated model: $\hat{w} = 52{,}912.77s^{-1.2060}l^{-1.3988}$. Using Minitab again, we find a 95% prediction interval for ln(w), when ln(s) = ln(60) and ln(l) = ln(6000), to be [2.176, 4.688]. Taking antilogs of the endpoints of this interval gives the 95% prediction interval for w: [8.81, 108.64].

190

57.   (a)   Since the data is paired (each subject is measured by both methods) we can conduct a two-sided paired t test of the null hypotheses H_0: $\mu_{BOD} - \mu_{HW} = 0$ so see if there is, on average, a significant difference between the two methods. The calculated t statistic for the paired t test of this data is $t = -3.54$, which is significantly different from 0 (assuming a significant level of $\alpha = .05$, $\pm t_{\alpha/2} = \pm t_{.025}$ (d.f. = n-1 = 19) = 2.093). Therefore, it appears that the average measurements given by the two methods are different.

       (b)   Regressing HW on BOD, we obtain the estimated regression line:
             HW = 4.788 + .743 BOD which has an r^2 value of .753.

59.   (a)   This is counted data, so we can compare the proportions of concussions among the three groups by using a chi-square test of the hypothesis H_0: $p_S = p_{NS} = p_C$. Arranging the data in a contingency table:

	soccer	non soccer	control
concussion:	45	28	8
no concussion:	46	68	45
totals:	91	96	45

             The χ^2 statistic associated with this table is $\chi^2 = 19.184$. The P-value of 19.184 (based on 2 d.f.) is less than .001 (from Table VII). Using a significance level of $\alpha = .05$, we can then conclude that there are significant differences between the three proportions. In particular, soccer players do suffer higher concussion rates than do the other two groups.

       (b)   The hypothesis test of H_0:$\rho = 0$ (no correlation) versus H_a: $\rho \neq 0$ can be conducted by calculating the test statistic $t = r\sqrt{\dfrac{n-2}{1-r^2}} = -.220\sqrt{\dfrac{91-2}{1-(.220)^2}} = -2.128$ and

comparing it to $\pm t_{\alpha/2}$ (d.f. = n-2 = 91-2 = 89). Using $\alpha = .05$, $\pm t_{.025}$ (89 d.f.) ≈ 1.99, so we can conclude that the correlation does differ from 0. That is, there is a significant negative correlation between the number of seasons played and memory score.

(c) These are independent samples, so we can conduct an independent samples t-test of the hypotheses H_0: $\mu_1-\mu_2 = 0$ versus H_a: $\mu_1-\mu_2 \neq 0$. The resulting t statistic is t = -.947, which is not significant at $\alpha = .05$, so this data does not show that there are any significant differences in test scores between the two groups.

(d) With this data we can conduct a one-way ANOVA test of H_0: $\mu_S = \mu_{NS} = \mu_C$. For this data, SSE = 124.2873 and SSTr = 3.4652 so the F statistic = MSTr / MSE = [SSTr/(3-1)] / [SSE/(237)] = 3.303. The critical F value associated with (2 df, 237 df) and $\alpha = .05$ is $F_{.05} \approx 3.03$. Since F = 3.303 > $F_{.05}$ we can conclude that there are differences between the numbers of non-soccer concussions.